OXFORD POPU

A Bid for Fortune, or

Guy Newell Boothby was born in 1867 in Adelaide, the son of an Australian politician. He was educated in England, and developed early the sense that he belonged (like his hero Dick Hatteras) to two countries. In 1883, aged 15, Boothby returned to Australia. He remained there eleven years, for most of the time employed as an assistant to the mayor of Adelaide. During this period he tried his hand at some plays for the stage, which proved unsuccessful. In 1891 Boothby undertook an arduous cross-continental trip across Australia, which would later furnish colour for his fiction. In 1894 he returned to England, where he threw himself into writing novels, principally for the new magazine-reading public generated by the successful launch of George Newnes's *Strand Magazine* in 1891. Almost immediately, Boothby had a huge success with *A Bid for Fortune, or Dr Nikola's Vendetta* (it was to remain his most reprinted and imitated work of fiction). Over the eleven years that remained to him, Boothby turned out a mass of romances in a variety of genres: detective stories, science fiction, nautical tales, Ruritanian fantasies, historical romances, outdoor adventure stories (some of them set in Australia—a feature which has led to a minor revival of interest in his work in that country in recent years). In ten years, Boothby produced fifty-three full-length novels under his own name, and probably others pseudonymously. He died prematurely from influenza at Boscombe in 1905, aged 38. Enriched by his fiction (which was extremely popular with readers of the late Victorian, early Edwardian period, although reviewers regularly complained that he was writing too much, too fast), Boothby was also an amateur farmer and dog-breeder.

John Sutherland is Lord Northcliffe Professor of Modern English Literature at University College London.

A Bid for Fortune

OR

Dr Nikola's Vendetta

Guy Boothby

Introduced by

John Sutherland

Oxford   New York
OXFORD UNIVERSITY PRESS
1996

Oxford University Press, Walton Street, Oxford OX2 6DP
Oxford New York
Athens Auckland Bangkok Bombay
Calcutta Cape Town Dar es Salaam Delhi
Florence Hong Kong Istanbul Karachi
Kuala Lumpur Madras Madrid Melbourne
Mexico City Nairobi Paris Singapore
Taipei Tokyo Toronto
and associated companies in
Berlin Ibadan

Oxford is a trade mark of Oxford University Press

Editional matter © John Sutherland 1996
First published as an Oxford paperback 1996

All rights reserved. No part of this publication may be reproduced,
stored in a retrieval system, or transmitted, in any form or by any means,
without the prior permission in writing of Oxford University Press.
Within the UK, exceptions are allowed in respect of any fair dealing for the
purpose of research or private study, or criticism or review, as permitted
under the Copyright, Designs and Patents Act, 1988, or in the case of
reprographic reproduction in accordance with the terms of the licences
issued by the Copyright Licensing Agency. Enquiries concerning
reproduction outside these terms and in other countries should be
sent to the Rights Department, Oxford University Press,
at the address above

This book is sold subject to the condition that it shall not, by way
of trade or otherwise, be lent, re-sold, hired out or otherwise circulated
without the publisher's prior consent in any form of binding or cover
other than that in which it is published and without a similar condition
including this condition being imposed on the subsequent purchaser

British Library Cataloguing in Publication Data
Data available

Library of Congress Cataloging in Publication Data
Data available
ISBN 0-19-283240-9

1 3 5 7 9 10 8 6 4 2

Typeset by Best-set Typesetter Ltd., Hong Kong
Printed in Great Britain by
Biddles Ltd
Guildford and King's Lynn.

OXFORD POPULAR FICTION

General Editor   Professor David Trotter
Associate Editor   Professor John Sutherland
Department of English, University College London

Amongst the many works of fiction that have become bestsellers and have then sunk into oblivion a significant number live on in popular consciousness, achieving almost folkloric status. Such books possess, as George Orwell observed, 'native grace' and have often articulated the collective aspirations and anxieties of their time more directly than so-called serious literature.

The aim of the Oxford Popular Fiction series is to introduce, or reintroduce, some of the most influential literary myth-makers of the last 150 years—bestselling works of British and American fiction that have helped define a new style or genre and that continue to resonate in popular memory. From crime and historical fiction to romance, adventure, and social comedy, the series will build up into a library of books that lie at the heart of British and American popular culture.

CONTENTS

INTRODUCTION	ix
A NOTE ON THE TEXT	xxi
FURTHER READING	xxii

PART I

	Prologue: Dr Nikola	3
I	I determine to take a Holiday	12
II	London	28
III	I Visit my Relations	39
IV	I Save an Important Life	52
V	Mystery	59
VI	I meet Dr Nikola again	71
VII	Port Said, and what Befell us There	86
VIII	Our Imprisonment and Attempt at Escape	98
IX	Dr Nikola permits us a Free Passage	109

PART II

I	We reach Australia, and the Result	123
II	On the Trail	135
III	Lord Beckenham's Story	149
IV	Following up a Clue	166
V	The Islands and what we found there	189
VI	Conclusion	203

INTRODUCTION

Imagine a line-up of the master criminals of 1890s popular fiction: Professor Moriarty, Svengali, H. G. Wells's Martians, Count Dracula, Dr Nikola, Prince Lucio (alias Satan).[1] What do these fiends in human (and inhuman) shape have in common? As any detective (or true-born Englishman) will at once observe, they all have names ending in vowel sounds—which English surnames never do. They are patently aliens: from Hell (via Italy), the Red Planet, Transylvania, Mittel-Europa, Ireland.[2] We never know by what names the Martians call themselves, except that their one identifiable vocalisation, 'Ulla, Ulla', does not roll easily off the Anglo-Saxon tongue. And it ends in a vowel.

What else do they have in common, apart from blatant un-Englishness? Most of them affect elaborately formal dress. Dracula, Nikola, Lucio, and Svengali (whenever Trilby sings under his sinister ministrations at the opera) are typically clad in capes, top hats, white tie, and tails—toff's gear (or, possibly, the garb of a foreign head-waiter). The Martians could, of course, hardly invade a neighbouring planet in top hats. But they crash around the English countryside in the smoke-spewing horseless carriages ('fighting machines') which, in 1897, were associated with the filthy rich.

Three of the villains are mesmerists, possessed of hypnotic powers which verge on the magical: Nikola, Svengali, and Satan. In his 'singular interview' with Moriarty in Baker Street (the prelude to their epic struggle at the Reichenbach Falls) it is clear that Moriarty is attempting to bring the great detective under his influence. 'His

---

[1] Professor James Moriarty features in 'The Final Problem' in Arthur Conan Doyle's *The Adventures of Sherlock Holmes* (1892); Svengali in George Du Maurier's *Trilby* (1894); Satan (Prince Lucio Rimanez) in Marie Corelli's *The Sorrows of Satan* (1895); Count Dracula in Bram Stoker's *Dracula* (1897); the Martians in Wells's *The War of the Worlds* (1897). All, including *A Bid for Fortune* (1895), are now available in World's Classics, or Popular Classics, volumes.

[2] Holmesians accept that Moriarty is 'of Irish extraction'—although it is recorded that he was brought up in the West of England. See 'The Napoleon of Crime: Prolegomena to a Memoir of Professor James Moriarty, Sc.D', by Edgar W. Smith, BSI, *The Baker Street Reader*, ed. Philip A. Shreffler (Westport: Conn., 1984), 79–88.

face', as Holmes recalls to Watson, 'protrudes forward, and is for ever slowly oscillating from side to side in a curiously reptilian fashion.' Luckily, Moriarty's passes fail to mesmerize his opponent although, on the only recorded occasion in his life, Holmes admits himself to be a frightened man.

Hypnotism allies itself with a Macchiavellian skill in orchestration of villainy behind the scenes. Nikola, Satan, Svengali, and Moriarty are masters of the long-term, sinister conspiracy against the English race. The Martians, we understand, have spent eons cunningly planning their arrival on Horsell Common. So too Dracula, with his fifty coffins secreted around London, intends to infect the whole country. These are enemies with deep-laid, long-meditated, and, above all, fiendishly clever plans. 'Moriarty is the Napoleon of crime,' Holmes informs Watson, 'He is the organizer of half that is evil and of nearly all that is undetected in this great city. He is a genius, a philosopher, an abstract thinker. He has a brain of the first order. He sits motionless, like a spider in the centre of its web, but that web has a thousand radiations, and he knows well every quiver of them. He does little himself. He only plans.' As Trilby says of her evil genius, Svengali, 'he's a rum 'un, ain't he? ... He reminds me of a big hungry spider and makes me feel like a fly!'

Like spiders (or Nikola's massive black cat, his only confidant) these arch-villains are predatory. Two of them actually consume English blood as their staple diet: Dracula and the Martians (oddly enough, they attack England in the same year, 1897, and could—by some fictional miscegenation—meet each other in the kind of 'Godzilla versus King Kong' encounter beloved of Japanese filmmakers). Svengali is vampiric in another way. All are sexually ambiguous in their appearance. They are variously epicene (Nikola, Moriarty) or degenerate (Svengali, Dracula) with oiled 'jet black' hair, sallow complexions, and strikingly aquiline profiles (Lucio, Nikola, Dracula, Svengali). The Martians have evolved a stage further and have transcended sex altogether, merging male and female into one androgynous unity. The sexual practices of these 'men' are, one guesses, wholly decadent. The honest English gorge rises at the thought of what Svengali does to his 'Drilpy' ('La Svengali' as the opera world knows her) in bed. Dracula gives ingenuous Englishmen like Jonathan Harker love-bites on the neck. The last we see of Lucio is arm-in-arm with a cabinet minister. Moriarty and Nikola have private sex lives into which we are never admitted, but at whose refined depravity we can only too easily guess. Nikola's effeminized name speaks volumes to English ears, and the description we have of him suggests that he may have wandered across to the *Windsor Magazine* ('an illustrated magazine for men and women' as it defiantly proclaimed itself) from the Beardsley–Wildean world of the *Yellow Book* (founded the year before, 1894, for something less than men, as Dick Hatteras would think). Nikola has 'long white fingers'; he is 'tall and slim, but exquisitely formed, and plainly the possessor of enormous strength'. In age, 'he might have been anything from eight and twenty to forty; in reality he was thirty-three'. Ambiguous again.

Two of the villains have aristocratic titles: Prince Lucio and Count Dracula. Two have academic titles: our Dr Nikola, and Professor Moriarty. Moriarty has actually taught at a university (a small provincial university, as Holmes rather snobbishly notes) and at 21 he published a treatise on the binomial theorem which has made his name in Europe—but not in England, we understand, where scientists are regarded with wholesome contempt. The Martians, who have evolved into pure cerebral entities, are even cleverer scientists than Moriarty. Svengali, although he has no handle to his one name was the most gifted pianist of his generation at the conservatory in Leipsig, where he studied, and whose lessons have apparently done him no good at all. On his part, Dr Nikola, we gather, is a virtuoso biologist with a pronounced interest in pharmaceuticals (in the sequel, *Dr Nikola's Experiment*, 1899, we discover that he has developed a drug which can extend life indefinitely; unfortunately, it is stolen by a villainous one-eared Chinaman setting off one of the chase-and-detection plots in which Boothby excelled).

The 1890s were an era when overdeveloped brains were automatically associated with racial decline. Max Nordau's book *Degeneration* was the great non-fiction bestseller of 1895. The English antidote to dangerous intellectualism and excessive sensibility was bluff, hearty, tweedy, philistinism. The hero's introduction of himself to the reader at the beginning of Chapter 1 of *A Bid for Fortune* is archetypal:

First and foremost, my name, age, description, and occupation, as they say in the *Police Gazette*. Richard Hatteras, at your service, commonly called Dick, of Thursday Island, North Queensland, pearler, copra merchant, *béche-de-mer* and tortoise-shell dealer, and South Sea trader generally. Eight-and-twenty years of age, neither particularly good-looking nor, if some people are to be believed, particularly amiable, six feet two in my stockings, and forty-six inches round the chest; strong as a Hakodate wrestler, and perfectly willing at any moment to pay ten pounds sterling to the man who can put me on my back.

Hatteras is not one to bother what little brains God has given him with binomial theorems or the kind of high-falutin stuff learned in Leipsig conservatories. But he is very healthy, and his 'stock' is excellent. And, as it turns out, he has an unusually good yarn to tell— one in which he features both as an intrepid hero and an ideal Briton. Ever since the 1840s and Carlyle's *Past and Present*, the big question for the English was how to regenerate their aristocracy—an echelon of the population which (with its access to luxury and its propensity for intermarriage) was forever sliding down the slippery slope to degeneration. The solution in *A Bid for Fortune* is to take the enfeebled youth, Lord Beckenham, to sea, throw away his books and Jermyn Street wardrobe, let him feel the brine on his face, clad him in seaman's canvas, teach him to use his fists, and generally make a man of him. (Rudyard Kipling made a novel on the same theme with *Captains Courageous*, the following year). Of course, Hatteras is no genius. Nikola finds it almost too easy to dupe him on the train ride down to Plymouth. But, at the end of the day, Dick's pluck, good heart, and manliness win through (although Nikola lives to fight on). Muscularity, we deduce, is worth more than coronets, and infinitely more than brains.

Like other villains of the 1890s, Dr Nikola can be seen to originate in British immigration phobias of the decade. Following state-sponsored pogroms in Russia and Poland, England—more specifically the East End of London—was being swamped by foreigners. Jews were particularly resented, for their religious difference and, often, their cleverness and business skill (this is a main theme in Mrs Humphry Ward's novel, *Sir George Tressady*, 1896). Fiction was powerful in fomenting prejudice but equally effective in allaying domestic alarm about newcomers. The Jewish community realized this, and Israel Zangwill was commissioned by the Jewish Publication Society to write his affectionate saga of life in Stepney, *Children of the Ghetto*, in 1892. The aim was to create a 'Jewish *Robert Elsmere*' (alluding to Mrs Humphry Ward's huge bestseller of 1888) which would present the 'peculiar people' in a sympathetic light to their English hosts.

A Bid for Fortune is by no means anti-Semitic (particularly in comparison with Du Maurier's odious depiction of Svengali in *Trilby*). But there is a strong hint that Nikola is Jewish, in the stress on his Mediterranean complexion, the jet black hair, and the slender, finely formed face, with its pronounced features. The form of the name indicates Russia—'Nikola', rather than 'Nicole', 'Nicolas', 'Niccolà', or 'Niklaus'. There is a clear allusion to Tsar Nicholas ('Nikolai') II, who succeeded his father in November 1894 and was enthroned in June 1895 (i.e. while Boothby's novel was appearing as a serial). On another front, the Russians were objects of particular terror in the 1890s, following anarchist outrages and assassinations in European cities. Dr Nikola, however, remains shrouded in racial and national vagueness. It is plausible to find a literary pedigree for him, going back through Wilkie Collins's Neapolitan Napoleon of Crime, Count Fosco, in *The Woman in White* (1860, Fosco's beloved white mice seem similarly to have suggested Nikola's black cat), and—delving to the roots—Montoni, the arch-Italianate, and omnipotent, villain in Mrs Radcliffe's *The Mysteries of Udolpho* (1794). Guy Boothby is not a novelist given to subtle effects. But his refusal to uncover his arch-villain's background is one of his finer touches.

To read *A Bid for Fortune* is to be plunged back into the popular reading experience of the 1890s—its helter-skelter excitements, its addiction to pace, adventure, suspense, and simple escape–capture–chase episodes. For the late twentieth-century reader *A Bid for Fortune* resembles nothing so much as the novelty film narratives, which were entrancing mass audiences in the 1890s and early 1900s. Early viewers wanted little more from 'motion pictures' than 'motion'. All that matters in *The Great Train Robbery* (1903) is the exhilarating sense of movement. Plot, logic, character—the apparatus which the novel had developed over three hundred years—are of secondary importance. All that mattered on the screen was the rushing velocity of the train, the galloping hooves, the shoot-out finale. Similarly, *A Bid for Fortune* is a novel built on speed, pace, and climax upon climax. As the *Athenaeum* noted (in a tart review of *A Bid for Fortune* in December 1895), Boothby does not connect his narrative, he simply 'piles it up'. The title, with its awkwardly latched-on subtitle (*A Bid for Fortune: or, Dr Nikola's Vendetta*) strongly suggests that Boothby may have begun with a quite different plot in his mind. What the significance of 'bid for fortune' is to the finally written narrative is baffling. 'Grab for Power' might make better sense. But such is the zest of Boothby's performance that anyone reading forgives the anomaly.

Well-titled or mis-titled, about one thing Boothby was quite clear when he projected his first (of three) novels for 1895. *A Bid for Fortune* was written primarily as an illustrated serial. The vehicle which Boothby had in mind for his story was the glossy, fiction-carrying magazine (later called 'slicks' in America), of which the pioneer was George Newnes's *Strand Magazine*, launched in 1891. In return for a modest 6*d.* the *Strand* offered 112 lavishly illustrated pages. The contents were extravagantly miscellaneous on the 'Tit-Bits' model which Newnes had introduced with spectacular success in the 1880s. For the outlay of their tanner, purchasers of the *Strand* might have, in a single issue, serials by Stanley Weyman and Grant Allen, a 'Sherlock Holmes' adventure, a 'case' involving Arthur Morrison's 'Investigator' (literature's first 'Private Eye') Martin Hewitt, a 'Raffles' ('Gentleman Cracksman') story, and a profusion of pictorial (and increasingly, after the early 1890s, photographic) features on the aristocracy, faraway places, and current events. In fiction, the *Strand* was a crucible in which English 'genre' fiction (notably the twentieth-century detective story) was formed.

The *Strand*'s sales were phenomenal, reaching half-a-million a month. It inspired rivals which slavishly copied Newnes's price, formula, and format. The most successful were *Pearson's Magazine* launched in 1896 by C. Arthur Pearson (H. G. Wells was an early contributor) and the *Windsor Magazine*, founded by Ward, Lock, and Bowden, in January 1895. The *Windsor*'s star writer (their Cònan Doyle) was Guy Boothby and one of its earliest big hits was *A Bid for Fortune: or, Dr Nikola's Vendetta*, which ran from January to August in the magazine's first year of issue alongside Stanley Weyman's nautical romp, *The Grey Lady* (one of that now-forgotten author's best efforts). Presumptuously, the *Windsor Magazine* advertised itself as 'the biggest and best sixpennyworth ever published', although its sales were probably nowhere near those of the *Strand* at its zenith. The *Windsor Magazine* featured on its cover a picturesque view of the royal castle, and—in its non-fiction department—specialized in the kind of 'royal human interest story' that was to become a staple in later twentieth-century tabloids. (The second issue of *A Bid for Fortune* in the magazine, for instance, was accompanied by a long article on Queen Victoria's tutors and governesses.) As with the *Strand*, illustration was profuse. *A Bid for Fortune*, for instance, had designs by no less than three artists (who did not entirely agree on what Nikola looked like, as reviewers complained). With fifty illustrations (in the favoured neo-photographic, realistic style favoured by the *Strand* and its imitators) for its 50,000 words of text, reading *A Bid for Fortune* was a distracting experience—there was so much to look at that it was sometimes hard to take in the words. Ward, Lock, and Bowden also brought out the 6*s*. library and cheap 6*d*. issues of the book versions of Boothby's stories. All were illustrated.

Fiction in the *Strand* and its imitators catered predominantly for the huge new readership recruited into literacy by the 1870 Universal Education Act. This meant, as a horrified Henry James put it, 'monstrous multiplications'. There was now a reading public of millions. They were not sophisticated, but neither were they sub- or semi-literate (as the large newly urbanized public for sub-Dickensian penny serials had been fifty years before).[3] Above all, they were young readers: they craved action, romance, thrills, fantasy, suspense, excitement. The new generation read fast, and they liked a good turn of speed in their narratives. They consumed on a large scale and encouraged their favourite authors to write voluminously. Boothby, who was among the fastest and most voluminous of his profession, turned out fifty novels in ten years—before dying prematurely in the 1905 influenza epidemic. His novels (apart from some interesting early experiments in Australian themes) were formulaic in the extreme, with no less than four improvised—and somewhat inferior—follow-ups to his first 'Nikola' tale, a creation who proved almost as hard for his creator to kill off as Holmes was for Conan Doyle.

With up to five titles a year flowing from his pen, Boothby was occasionally repetitive and, not infrequently, light-fingered where the work of others was concerned. Clearly, Nikola owes much to Moriarty. Nikola's mesmeric tricks are lifted from Svengali's similar pranks in *Trilby*. Readers conversant with 1890s literature will catch unmistakable echoes of Stevenson's *Beach of Falesà* (in the final duel to the death on the South Seas Island, and in John Wiltshire's bluff

---

[3] See Louis James, *Fiction for the Working Man, 1830–50* (Oxford, 1963).

monologue, which is very close to that of Dick Hatteras, who has been a trader in exactly the same waters as the hero of Stevenson's 1892 romance). The elegant Nikola's resorting, when not in the gilded saloons of the West End, to 'the Green Sailor public-house, East India Dock Road' recalls Dorian Gray's slumming (also dressed to the nines) in Limehouse in Oscar Wilde's 1892 novel.

Boothby's most flagrant borrowing is found in Chapter 9, when Dick and Lord Beckenham ingeniously escape from the Port Said cellar where they have been chained for some weeks by the neck (Nikola's sinister reason for this elaborate form of confinement, rather than simply having his victims knifed and dumped in the harbour, is left to the reader's imagination). The heroes break down a door of their prison only to find themselves in Dr Nikola's laboratory. Hideous specimens in bottles surround them. 'Words would fail me if I tried to give you a true and accurate description of it', Dick rather too candidly tells the reader. Hundreds of animal cadavers litter the room. Chained by Nikola's bench is a 'native of Northern India, if one might judge by his dress and complexion'. He has a head three times too big for his body, supported by an iron tripod. Alongside this monster is 'a Burmese monkey-boy ... chained to the wall ... chattering and scratching for all the world like a monkey at a zoo'. Meanwhile at the operating table itself Nikola—apparently unconcerned that his prisoners have escaped—calmly continues 'dissecting an animal strangely resembling a monkey' (the fact that it is *not* a monkey can only mean—horrors!—that it is a human fetus) watched all the time by his inseparable companion, the gigantic black cat Apollyon, and 'an albino dwarf, scarcely more than two feet eight inches high'.

Anyone reading this vivisectionist nightmare will recall Wells's *Island of Dr Moreau*. Clearly, like his fellow physician, Nikola is experimenting with the surgical transmutation of animals into humans. And since Wells's scientific romance was not published until 1896, it might seem that Boothby (whose *A Bid for Fortune* was wholly in print by September 1895) has priority here. He does not. Although *The Island of Dr Moreau* was not serialized, chapter 14 ('Dr Moreau Explains' in the novel) was published as an article called 'The Limits of Individual Plasticity' in the *Saturday Review*, 19 January 1895. The extracted chapter outlines Moreau's hideous experiments, leading up to his vivisectionist operation on a gorilla from whom, 'working with infinite care and mastering difficulty after difficulty, I made my first man'. It seems clear that Boothby read the piece, and happily incorporated it into his work in progress. It is, of course, a somewhat unlikely episode, if one thinks for a moment. Why would Nikola, who has just come from a long stay in Brazil (where he has nefarious business in the diamond fields) and is just about to set off for Australia (where he plans a fiendish personation of Lord Beckenham) bother to set up a laboratory in Port Said, acquiring a Burmese monkey-boy (not easily come by in Egypt, one would think) for his experiments and hiring an albino dwarf as his lab-assistant? But Boothby liked Wells's *grand guignol*, so in it went, to create one of the novel's most effective climaxes.

A Bid for Fortune was published in eight parts, between January and August 1895, the instalments in the *Windsor Magazine* comprising on average two chapters of the narrative. Each part contained a fight, a daring escape or chase, and usually some romantic interest. The last is never protracted: in the first episode Dick rescues Phyllis from hooligans in Sidney, has a shipboard romance with her, proposes in his blunt-spoken way, is accepted by the coy maiden, and banished by her irate father in London—all in fifteen pages. Every number of *A Bid for Fortune*, except the last, ends with a baited hook of the 'little did I dream of the unspeakable horrors that awaited me when I threw open the door' variety. The story moves from one far-flung corner of the earth to the other with dizzying speed: one minute we are in England, the next in Australia, a stopover in Egypt intervenes, then the reader is rushed into a tremendous final episode scudding through the foam of the South Seas. The story is frequently in such a hurry to get to its next destination, that the author is obliged to apologize for not being able to do his job as conscientiously as he would like. He cannot, for instance, find time to describe his villain on his first appearance in the novel: 'It would take more time than I can spare the subject', he tells us, 'to give you an adequate and inclusive description of the person who entered the room at that moment.' Describe him yourself, in other words; the author of *A Bid for Fortune* has a story to get on with.

Boothby is never one to linger over his narrative duties. Three years later, in 1898, Henry James was to elaborate his theory of the gradual intensification of effect in *The Turn of the Screw*. On his part, Boothby believed that if you moved fast enough, the reader would not have time to ask questions. The pen could be quicker than the eye. With Boothby, it is less the slow turning of the screw than the blur of a propeller at full speed. 'He never allows the interest to drop from first page to last', observed *The Times* of *A Bid for Fortune*. A cost in logic is paid for the never-flagging 'interest'. Boothby is the grandmaster of the loose end—the 'no time to explain that now' gambit. There is a lot that is left unexplained in *A Bid for Fortune*, some of it central to the plot. With an effrontery which verges on the magnificent, Boothby withholds from us what is the significance of the mysterious talisman for which Nikola is prepared to abduct the heroine and hold her hostage in the South Seas. Why all this fuss about China Pete's 'queer little wooden stick about three and a half inches long, made of some heavy timber, and covered all over with Chinese inscriptions'? Something about 'ten million pounds sterling'? Slavering with expectation, the reader turns eagerly to the last page only to be met by a denouement in which nothing is undone. 'What gigantic *coup* Nikola intends to accomplish with the little Chinese stick . . . is beyond my power to tell', Hatteras coolly informs us: 'I am only too thankful, however, that I am able to say that I am not the least concerned in it.' Nor should readers hope for explanation in any of *A Bid for Fortune*'s sequels.

In order to appreciate *A Bid for Fortune* in the spirit in which the tale is offered one must devour it in a breathless rush and ask no questions afterwards. Looked at with the slightest afterthought the story falls to pieces. The plot is compounded of gross improbabilities, held together by grosser improbabilities. It all works, so long as one suspends any unfriendly scepticism. Improbability pervades even the best thing in the narrative—the prelude at 'the new Imperial Restaurant on the Thames Embankment'.[4] This, interestingly enough, is the only scene in the whole protracted Nikola sequence of novels which is given in third-person narrative, rather than autobiographically by some hearty, but not very eloquent hero. It is early evening and dinner is in prospect. The manager of the Imperial, Mr McPherson, has received three months earlier a letter from Cuyaba, 'a

---

[4] Boothby clearly indicates the Savoy Hotel, which fronts the Strand and backs on to the Embankment. Oddly enough there was a new Imperial Hotel in 1895 being built on the east side of Russell Square (an even newer Imperial Hotel now stands in its place).

town almost on the western or Bolivian border of Brazil—considerably connected with the famous Brazilian diamond fields'. He is to prepare his finest private supper room for a meeting at precisely 8 o'clock that evening. In addition to the mysterious host, there will be three guests: one from Hang-chow, one from Bloemfontein, and the third from England. All this is detailed in the letter—and seems rather more than is necessary to tell a hotel manager. There is to be 'no electric light', but 'candles with red shades'. A porcelain saucer, and a small jug of new milk are to be provided (for the host's feline companion, we discover). Dr Nikola signs his letter of instruction with his own name.

It is all very atmospheric and serves its purpose splendidly. No-one reading the sinister opening episode in the Imperial Hotel is going to stop reading *A Bid for Fortune*. But since Nikola is hatching a conspiracy to defraud one of the most prominent noblemen in England of vast amounts of money it might be thought indiscreet to gather his accomplices for their briefing session in quite such a flamboyant fashion. His drinking den in the East India Road would seem more appropriate. Whatever else he forgets, the amazing Dr Nikola, his black cat, and his red-shaded candles are not going to slip Mr McPherson's mind when Scotland Yard comes enquiring. Moreover, Nikola's scheme itself is no masterpiece of crime. Why, if he wants to revenge himself on an English marquess, does he arrange to have his victim's son impersonated in Australia, without taking the obvious precaution of doing away with the real Lord Beckenham on one of those many occasions that the young aristocrat is entirely in his power? For a vivisectionist happy to chop dark-skinned natives into little pieces with his scalpel, Nikola is strangely reluctant to hurt white men. And why, having apparently wanted nothing but money from the Marquess of Glenbarth, does Nikola suddenly change his mind and decide that what he really wants is China Pete's stick from the Colonial Secretary, Mr Wetherell? If Boothby knows the answers they are not to be found in *A Bid for Fortune*.

A Bid for Fortune has to be enjoyed for what it is. If ever a yarn rattled, this one does. Boothby never quite succeeded in pulling the trick off again. The novel's four dashed-off successors—*Doctor Nikola* (1896), *The Lust of Hate* (1898), *Dr Nikola's Experiment* (1899), and *Farewell, Nikola* (1901)—make disappointing reading after the zestiness of the first in the series. Dick Hatteras ('Lord Hatteras', as he finally becomes, in a Tichborne-like twist)[5] is reintroduced as the narrator of the last volume. It is possible that, like Doyle and Holmes, Boothby might have been prevailed on to resurrect Nikola, had the author not died in 1905. But, unlike the great detective, whose formula was to prove inexhaustible, it is extremely unlikely that the wonderful exuberance of *A Bid for Fortune* could have been recovered.[6]

Boothby's novel did very well on its first appearance, and enjoyed the late-Victorian equivalent of Warhol's fifteen minutes of world fame. But, such as the chances of popular fiction, it has not lasted as well as any of the other romances of glamorous villainy mentioned in the first paragraph above. Even Marie Corelli has a higher degree of name recognition among general readers in the 1990s than Guy Newell Boothby. Nor has Dr Nikola entered folklore as has Moriarty, or Dorian Grey—let alone Dracula. None the less, *A Bid for Fortune* can be seen to have had famous progeny in popular narrative of the twentieth century. In British and American fiction the villainous doctor, following the trail of Dr Jekyll and Dr Moreau, has been stereotyped into the mad scientist—a line which climaxes majestically in Stanley Kubrick's and Terry Southern's Dr Strangelove. A different stereotype emerged in two immensely influential German expressionist films of the silent period: Robert Wiene's *Das Cabinett Des Doktor Caligari* ('The Cabinet of Dr Caligari', 1919) and Fritz Lang's *Doktor Mabuse Der Spieler* ('Dr Mabuse, the Gambler', 1922). Doctors Caligari and Mabuse are hypnotists, and master-criminals with megalomaniac ambitions to take over the world. The theme was, we are told by film historians, popular in postwar Germany. It seems quite clearly to derive from Boothby's dark, proto-fascistic, conception of Dr Nikola and his sinister bid for fortune.

[5] The Tichborne affair was a favourite theme of sensation novelists in the second half of the nineteenth century. In 1854 the heir to the Tichborne estate and title was drowned at sea. In 1865 an Australian butcher, called Arthur Orton, claimed to be the lost heir, Sir Roger Tichborne. After much publicity, the 'claimant' was judged to be an impostor, and sentenced to fourteen years' penal servitude in 1874. But the myth of British aristocrats turning up, like Dick Hatteras, in the Australian outback was firmly planted in the popular mind by the case.

[6] In *Farewell Nikola*, Hatteras is a husband of four years, happily married to Phyllis. The couple are on holiday in Venice, where their paths cross again with the sinister doctor. Nikola is revealed to be of aristocratic Venetian origin. A reformed character, he makes amends to Hatteras before disappearing to end his days as a Buddhist priest in the Himalayas.

A NOTE ON THE TEXT

A Bid for Fortune: or, Dr Nikola's Vendetta was serialized in the *Windsor Magazine*, January to September 1895, illustrated by Stanley L. Wood, Oscar Eckhardt, and T. S. C. Crowther. The instalments were prefaced with plot-to-date summaries, which were probably not supplied by Boothby. The one-volume, 6*s.* book version of the novel (with all the illustrations) was published by Ward, Lock, and Bowden (the *Windsor Magazine*'s publishers) in late November 1895. It is from this text that this edition is taken, although the running heads (which vary from page to page) have been dropped with repagination.

FURTHER READING

The only reliable source of information on Boothby which I know of is the *DNB* (Supplement, 1901–1911) entry, which has appended to it a list of obituary notices. What follows is as full a listing of Boothby's titles (novels and collections of short stories) as can be assembled from such sources as the British Library Catalogue and contemporary advertisements. The subjects and generic styles of the narratives are usually clear enough from the titles. One oddity is *The Phantom Stockman*, which was printed in shorthand.

Across the World for a Wife, 1896
The Beautiful White Devil, 1896
A Bid for Fortune, or Dr Nikola's Vendetta, 1895
A Bid for Freedom, 1904
Billy Binks, Hero, 1898
A Bride from the Sea, 1904
A Brighton Tragedy, 1905
Bushigrams, 1897
A Cabinet Secret, 1901
The Childerbridge Mystery, 1902
Connie Burt, 1903
A Consummate Scoundrel, 1904
The Countess Londa, 1902
A Crime of the Under-Seas, 1905
The Curse of the Snake, 1902
A Desperate Conspiracy, 1904
Doctor Nikola, 1896
Dr Nikola's Experiment, 1899
Farewell Nikola, 1901
The Fascination of the King, 1897
For Love of Her, 1905
In Spite of the Czar, 1905
In Strange Company, 1894
The Kidnapped President, 1902
The Lady of the Island, 1904
The League of Twelve, 1903
Long Live the King, 1900
A Lost Endeavour, 1914
Love Made Manifest, 1899
The Lust of Hate, 1898
A Maker of Nations, 1900
The Man of the Crag, 1907
The Marriage of Esther, 1895
A Millionaire's Love Story, 1901
My Australian Duchess, 1903
My Indian Queen, 1901
My Strangest Case, 1901
The Mystery of the Clasped Hands, 1901
An Ocean Secret, 1904
On the Wallaby, 1894
The Phantom Stockman, 1897
Pharos the Egyptian, 1899
A Prince of Swindlers, 1900
A Queer Affair, 1903
The Race of Life, 1906
The Red Rat's Daughter, 1899
A Royal Affair, 1906
A Sailor's Bride, 1900
Sheila McLeod, 1897
Stephen Witledges' Revenge, 1904
A Stolen Peer, 1906
A Two-Fold Inheritance, 1903
Uncle Joe's Legacy, 1902
The Woman of Death, 1900

A Bid for Fortune

OR

Dr Nikola's Vendetta

TO MY DEAR WIFE,
TO WHOM IT OWES SO MUCH,
I
DEDICATE THIS BOOK

PART I

PROLOGUE
Dr Nikola

The manager of the new Imperial Restaurant on the Thames Embankment went into his luxurious private office and shut the door. Having done so, he first scratched his chin reflectively, and then took a letter from the drawer in which it had reposed for more than two months and perused it carefully. Though he was not aware of it, this was the thirtieth time he had read it since breakfast that morning. And yet he was not a whit nearer understanding it than he had been at the beginning. He turned it over and scrutinized the back, where not a sign of writing was to be seen; he held it up to the window, as if he might hope to discover something from the watermark; but there was evidently nothing in either of these places of a nature calculated to set his troubled mind at rest. Then, though he had a clock upon his mantelpiece in good working order, he took a magnificent repeater watch from his waistcoat pocket and glanced at the dial; the hands stood at half-past seven. He immediately threw the letter on the table, and as he did so his anxiety found relief in words.

'It's really the most extraordinary affair I ever had to do with,' he remarked to the placid face of the clock above mentioned. 'And as I've been in the business just three-and-thirty years at eleven a.m. next Monday morning, I ought to know something about it. I only hope I've done right, that's all.'

As he spoke, the chief bookkeeper, who had the treble advantage of being tall, pretty, and just eight-and-twenty years of age, entered the room. She noticed the open letter and the look upon her chief's face, and her curiosity was proportionately excited.

'You seem worried, Mr McPherson,' she said tenderly, as she put down the papers she had brought in for his signature.

'You have just hit it, Miss O'Sullivan,' he answered, pushing them further on to the table. 'I am worried about many things, but particularly about this letter.'

He handed the epistle to her, and she, being desirous of impressing him with her business capabilities, read it with ostentatious care. But it was noticeable that when she reached the signature she too turned back to the beginning, and then deliberately read it over again. The manager rose, crossed to the mantelpiece, and rang for the head waiter. Having relieved his feelings in this way, he seated himself again at his writing-table, put on his glasses, and stared at his companion, while waiting for her to speak.

'It's very funny,' she said at length, seeing that she was expected to say something. 'Very funny indeed!'

'It's the most extraordinary communication I have ever received,' he replied with conviction. 'You see it is written from Cuyaba, Brazil. The date is three months ago to a day. Now I have taken the trouble to find out where and what Cuyaba is.'

He made this confession with an air of conscious pride, and having done so, laid himself back in his chair, stuck his thumbs into the arm-holes of his waistcoat, and looked at his fair subordinate for approval. Nor was he destined to be disappointed. He was a bachelor in possession of a snug income, and she, besides being a pretty woman, was a lady with a keen eye to the main chance.

'And where *is* Cuyaba?' she asked humbly.

'Cuyaba,' he replied, rolling his tongue with considerable relish round his unconscious mispronunciation of the name, 'is a town almost on the western or Bolivian border of Brazil. It is of moderate size, is situated on the banks of the river Cuyaba, and is considerably connected with the famous Brazilian Diamond Fields.'

'And does the writer of this letter live there?'

'I cannot say. He writes from there—that is enough for us.'

'And he orders dinner for four—here, in a private room overlooking the river, three months ahead—punctually at eight o'clock, gives you a list of the things he wants, and even arranges the decoration of the table. Says he has never seen either of his three friends before; that one of them hails from (here she consulted the letter again) Hang-chow, another from Bloemfontein, while the third resides, at present, in England. Each one is to present an ordinary visiting card with a red dot on it to the porter in the hall, and to be shown to the room at once. I don't understand it at all.'

The manager paused for a moment, and then said deliberately—

'Hang-chow is in China, Bloemfontein is in South Africa.'

'What a wonderful man you are, to be sure, Mr McPherson! I never can *think* how you manage to carry so much in your head.'

There spoke the true woman. And it was a move in the right direction, for the manager was susceptible to her gentle influence, as she had occasion to know.

At this juncture the head waiter appeared upon the scene, and took up a position just inside the doorway, as if he were afraid of injuring the carpet by coming further.

'Is No 22 ready, Williams?'

'Quite ready, sir. The wine is on the ice, and cook tells me he'll be ready to dish punctual to the moment.'

'The letter says, "no electric light; candles with red shades". Have you put on those shades I got this morning?'

'Just seen it done this very minute, sir.'

'And let me see, there was one other thing.' He took the letter from the chief bookkeeper's hand and glanced at it. 'Ah, yes, a porcelain saucer, and a small jug of new milk upon the mantelpiece. An extraordinary request, but has it been attended to?'

'I put it there myself, sir.'

'Who wait?'

'Jones, Edmunds, Brooks, and Tomkins.'

'Very good. Then I think that will do. Stay! You had better tell the hall porter to look out for three gentlemen presenting plain visiting cards with a little red spot on them. Let Brooks wait in the hall, and when they arrive tell him to show them straight up to the room.'

'It shall be done, sir.'

The head waiter left the room, and the manager stretched himself in his chair, yawned by way of showing his importance, and then said solemnly—

'I don't believe they'll any of them turn up; but if they do, this Dr Nikola, whoever he may be, won't be able to find fault with my arrangements.'

Then, leaving the dusty high road of Business, he and his companion wandered in the shady bridle-paths of Love—to the end that when the chief bookkeeper returned to her own department she had forgotten the strange dinner party about to take place upstairs, and was busily engaged upon a calculation as to how she would look in white satin and orange blossoms, and, that settled, fell to wondering whether it was true, as Miss Joyce, a subordinate, had been heard to declare, that the manager had once shown himself partial to a certain widow with reputed savings and a share in an extensive egg and dairy business.

At ten minutes to eight precisely a hansom drew up at the steps of the hotel. As soon as it stopped, an undersized gentleman, with a clean shaven countenance, a canonical corporation, and bow legs, dressed in a decidedly clerical garb, alighted. He paid and discharged his cabman, and then took from his ticket pocket an ordinary white visiting card, which he presented to the gold-laced individual who had opened the apron. The latter, having noted the red spot, called a waiter, and the reverend gentleman was immediately escorted upstairs.

Hardly had the attendant time to return to his station in the hall, before a second cab made its appearance, closely followed by a third. Out of the second jumped a tall, active, well-built man of about thirty years of age. He was dressed in evening dress of the latest fashion, and to conceal it from the vulgar gaze, wore a large Inverness cape of heavy texture. He also in his turn handed a white card to the porter, and, having done so, proceeded into the hall, followed by the occupant of the last cab, who had closely copied his example. This individual was also in evening dress, but it was of a different stamp. It was old-fashioned and had seen much use. The wearer, too, was taller than the ordinary run of men, while it was noticeable that his hair was snow-white, and that his face was deeply pitted with smallpox. After disposing of their hats and coats in an ante-room, they reached room No 22, where they found the gentleman in clerical costume pacing impatiently up and down.

Left alone, the tallest of the trio, who for want of a better title we may call the Best Dressed Man, took out his watch, and having glanced at it, looked at his companions.

'Gentlemen,' he said, with a slight American accent, 'it is three minutes to eight o'clock. My name is Eastover!'

'I'm glad to hear it, for I'm most uncommonly hungry,' said the next tallest, whom I have already described as being so marked by disease. 'My name is Prendergast!'

'We only wait for our friend and host,' remarked the clerical gentleman, as if he felt he ought to take a share in the conversation, and then, as if an afterthought had struck him, he continued, 'My name is Baxter!'

They shook hands all round with marked cordiality, seated themselves again, and took it in turns to examine the clock.

'Have you ever had the pleasure of meeting our host before?' asked Mr Baxter of Mr Prendergast.

'Never,' replied that gentleman, with a shake of his head. 'Perhaps Mr Eastover has been more fortunate?'

'Not I,' was the brief rejoinder. 'I've had to do with him off and on for longer than I care to reckon, but I've never set eyes on him up to date.'

'And where may he have been the first time you heard from him?'

'In Nashville, Tennessee,' said Eastover. 'After that, Tahupapa, New Zealand; after that, Papeete, in the Society Islands; then Pekin, China. And you?'

'First time, Brussels; second, Monte Video; third, Mandalay, and then the Gold Coast, Africa. It's your turn, Mr Baxter.'

The clergyman glanced at the timepiece. It was exactly eight o'clock.

'First time, Cabul, Afghanistan; second, Nijni Novgorod, Russia; third, Wilcannia, Darling River, Australia; fourth, Valparaiso, Chili; fifth, Nagasaki, Japan.'

'He is evidently a great traveller and a most mysterious person.'

'He is more than that,' said Eastover with conviction; 'he is late for dinner!'

Prendergast looked at his watch.

'That clock is two minutes fast. Hark, there goes Big Ben! Eight exactly.'

As he spoke the door was thrown open and a voice announced 'Dr Nikola.'

The three men sprang to their feet simultaneously, with exclamations of astonishment, as the man they had been discussing made his appearance.

It would take more time than I can spare the subject to give you an adequate and inclusive description of the person who entered the room at that moment. In stature he was slightly above the ordinary, his shoulders were broad, his limbs perfectly shaped and plainly muscular, but very slim. His head, wich was magnificently set upon his shoulders, was adorned with a profusion of glossy black hair; his face was destitute of beard or moustache, and was of oval shape and handsome moulding; while his skin was of a dark olive hue, a colour which harmonized well with his piercing black eyes and pearly teeth. His hands and feet were small, and the greatest dandy must have admitted that he was irreproachably dressed, with a neatness that bordered on the puritanical. In age he might have been anything from eight-and-twenty to forty; in reality he was thirty-three. He advanced into the room and walked with outstretched hand directly across to where Eastover was standing by the fireplace.

'Mr Eastover, I feel certain,' he said, fixing his glittering eyes upon the man he addressed, and allowing a curious smile to play upon his face.

'That is my name, Dr Nikola,' the other answered with evident surprise. 'But how on earth can you distinguish me from your other guests?'

'Ah! it would surprise you if you knew. And Mr Prendergast, and Mr Baxter. This is delightful; I hope I am not late. We had a collision in the Channel this morning, and I was almost afraid I might not be up to time. Dinner seems ready; shall we sit down to it?'

They seated themselves, and the meal commenced. The Imperial Restaurant has earned an enviable reputation for doing things well, and the dinner that night did not in any way detract from its lustre. But delightful as it all was, it was noticeable that the three guests paid more attention to their host than to his excellent menu. As they had said before his arrival, they had all had dealings with him for several years, but what those dealings were they were careful not to describe. It was more than possible that they hardly liked to remember them themselves.

When coffee had been served and the servants had withdrawn, Dr Nikola rose from the table, and went across to the massive sideboard. On it stood a basket of very curious shape and workmanship. This he opened, and as he did so, to the astonishment of his guests, an enormous cat, as black as his master's coat, leaped out on to the floor. The reason for the saucer and jug of milk became evident.

Seating himself at the table again, the host followed the example of his guests and lit a cigar, blowing a cloud of smoke luxuriously through his delicately chiselled nostrils. His eyes wandered round the cornice of the room, took in the pictures and decorations, and then came down to meet the faces of his companions. As they did so, the black cat, having finished its meal, sprang on to his shoulder to crouch there, watching the three men through the curling smoke drift with its green, blinking, fiendish eyes.

Dr Nikola smiled as he noticed the effect the animal had upon his guests.

'Now shall we get to business?' he said briskly.

The others almost simultaneously knocked the ashes off their cigars and brought themselves to attention. Dr Nikola's dainty, languid manner seemed to drop from him like a cloak, his eyes brightened, and his voice, when he spoke, was clean cut as chiselled silver.

'You are doubtless anxious to be informed why I summoned you from all parts of the globe to meet me here tonight? And it is very natural you should be. But then from what you know of me you should not be surprised at anything I do.'

His voice gradually dropped back into its old tone of gentle languor. He drew in a great breath of smoke and then sent it slowly out from his lips again. His eyes were half closed, and he drummed with one finger on the table edge. The cat looked through the smoke at the three men, and it seemed to them that he grew every moment larger and more ferocious. Presently his owner took him from his perch and seating him on his knee fell to stroking his fur, from head to tail, with his long slim fingers. It was as if he were drawing inspiration for some deadly mischief from the uncanny beast.

'To preface what I have to say to you, let me tell you that this is by far the most important business for which I have ever required your help. (Three slow strokes down the centre of the back and one round each ear.) When it first came into my mind I was at a loss who to trust in the matter. I thought of Vendon, but I found Vendon was dead. I thought of Brownlow, but Brownlow was no longer faithful. (Two strokes down the back and two on the throat.) Then bit by bit I remembered you. I was in Brazil at the time. So I sent for you. You came, and we meet here. So far so good.'

He rose and crossed over to the fireplace. As he went the cat crawled back to its original position on his shoulder. Then his voice changed once more to its former business-like tone.

'I am not going to tell you very much about it. But from what I do tell you, you will be able to gather a great deal and imagine the rest. To begin with, there is a man living in this world today who has done me a great and lasting injury. What that injury is is no concern of yours. You would not understand if I told you. So we'll leave that out of the question. He is immensely rich. His cheque for £300,000 would be honoured by his bank at any minute. Obviously he is a power. He has had reason to know that I am pitting my wits against his, and he flatters himself that so far he has got the better of me. That is because I am drawing him on. I am maturing a plan which will make him a poor and a very miserable man at one and the same time. If that scheme succeeds, and I am satisfied with the way you three men have performed the parts I shall call on you to play in it, I shall pay to each of you the sum of £10,000. If it doesn't succeed, then you will each receive a thousand and your expenses. Do you follow me?'

It was evident from their faces that they hung upon his every word.

'But, remember, I demand from you your whole and entire labour. While you are serving me you are mine body and soul. I know you are trustworthy. I have had good proof that you are—pardon the expression—unscrupulous, and I flatter myself you are silent. What is more, I shall tell you nothing beyond what is necessary for the carrying out of my scheme, so that you could not betray me if you would. Now for my plans!'

He sat down again and took a paper from his pocket. Having perused it, he turned to Eastover.

'You will leave at once—that is to say, by the boat on Wednesday—for Sydney. You will book your passage tomorrow morning, first thing, and join her in Plymouth. You will meet me tomorrow evening at an address I will send you and receive your final instructions. Goodnight.'

Seeing that he was expected to go, Eastover rose, shook hands, and left the room without a word. He was too astonished to hesitate or to say anything.

Nikola took another letter from his pocket and turned to Prendergast.

Prologue: Dr Nikola

'*You* will go down to Dover tonight, cross to Paris tomorrow morning, and leave this letter personally at the address you will find written on it. On Thursday, at half-past two precisely, you will deliver me an answer in the porch at Charing Cross. You will find sufficient money in that envelope to pay all your expenses. Now go!'

'At half-past two you shall have your answer. Good-night.'

'Good-night.'

When Prendergast had left the room, Dr Nikola lit another cigar and turned his attentions to Mr Baxter.

'Six months ago, Mr Baxter, I found for you a situation as tutor to the young Marquis of Beckenham. You still hold it, I suppose?'

'I do.'

'Is the Duke, the lad's father, well disposed towards you?'

'In every way. I have done my best to ingratiate myself with him. That was one of your instructions, if you will remember.'

'Yes, yes! But I was not certain that you would succeed. If the old man is anything like what he was when I last met him, he must still be a difficult person to deal with. Does the boy like you?'

'I hope so.'

'Have you brought me his photograph as I directed?'

'I have. Here it is.'

Baxter took a photograph from his pocket and handed it across the table.

'Good. You have done very well, Mr Baxter. I am pleased with you. Tomorrow morning you will go back to Yorkshire——'

'I beg your pardon, Bournemouth. His Grace owns a house near Bournemouth, which he occupies during the summer mouths.'

'Very well—then tomorrow morning you will go back to Bournemouth and continue to ingratiate yourself with father and son. You will also begin to implant in the boy's mind a desire for travel. Don't let him become aware that his desire has its source in you—but do not fail to foster it all you can. I will communicate with you further in a day or two. Now go.'

Baxter in his turn left the room. The door closed. Dr Nikola picked up the photograph and studied it carefully.

'The likeness is unmistakable—or it ought to be. My friend, my very dear friend, Wetherell, my toils are closing on you. My arrangements are perfecting themselves admirably. Presently when all is complete I shall press the lever, the machinery will be set in motion, and you will find yourself being slowly but surely ground into powder. Then you will hand over what I want, and be sorry you thought fit to baulk Dr Nikola!'

He rang the bell and ordered his bill. This duty discharged he placed the cat back in its prison, shut the lid, descended with the basket to the hall, and called a hansom. When he had closed the apron, the porter enquired to what address he should order the cabman to drive. Dr Nikola did not reply for a moment, then he said, as if he had been thinking something out:

'The Green Sailor public-house, East India Dock Road.'

CHAPTER I

I Determine to take a Holiday.—Sydney, and what Befel me There

First and foremost, my name, age, description, and occupation, as they say in the *Police Gazette*. Richard Hatteras, at your service, commonly called Dick, of Thursday Island, North Queensland, pearler, copra merchant, *béche-de-mer* and tortoise-shell dealer, and South Sea trader generally. Eight-and-twenty years of age, neither particularly good-looking nor, if some people are to be believed, particularly amiable, six feet two in my stockings, and forty-six inches round the chest; strong as a Hakodate wrestler, and perfectly willing at any moment to pay ten pounds sterling to the man who can put me on my back.

And big shame to me if I were not so strong, considering the free, open-air, devil-may-care life I've led. Why, I was doing man's work at an age when most boys are wondering when they're going to be taken out of knickerbockers. I'd been half round the world before I was fifteen, and had been wrecked twice and marooned once before my beard showed signs of sprouting. My father was an Englishman, not very much profit to himself, so he used to say, but of a kindly disposition, and the best husband to my mother, during their short married life, that any woman could possibly have desired. She, poor soul, died of fever in the Philippines the year I was born, and he went to the bottom in the schooner *Helen of Troy*, a degree west of the Line Islands, within six months of her decease; struck the tail end of a cyclone, it was thought, and went down, lock, stock, and barrel, leaving only one man to tell the tale. So I lost father and mother in the same twelve months, and that being so, when I put my cabbage-tree on my head it covered, as far as I knew, all my family in the world.

Any way you look at it, it's calculated to give you a turn, at fifteen years of age, to know that there's not a living soul on the face of God's globe that you can take by the hand and call relation. That old saying about 'blood being thicker than water' is a pretty true one, I reckon: friends may be kind—they were so to me—but after all they're not the same thing, nor can they be, as your own kith and kin.

However, I had to look my trouble in the face and stand up to it as a man should, and I suppose this kept me from brooding over my loss as much as I should otherwise have done. At any rate, ten days after the news reached me, I had shipped aboard the *Little Emily*, trading schooner, for Papeete, booked for five years among the islands, where I was to learn to water copra, to cook my balances, and to lay the foundation of the strange adventures that I am going to tell you about in this book.

After my time expired and I had served my Trading Company on half the mudbanks of the Pacific, I returned to Australia and went up inside the Great Barrier Reef to Somerset—the pearling station that had just come into existence on Cape York. They were good days there then, before all the new-fangled laws that now regulate the pearling trade had come into force; days when a man could do almost as he liked among the islands in those seas. I don't know how other folk liked it, but the life just suited me—so much so that when Somerset proved inconvenient and the settlement shifted across to Thursday, I went with it, and, what was more to the point, with money enough at my back to fit myself out with a brand new lugger and full crew, so that I could go pearling on my own account.

For many years I went at it head down, and this brings me up to four years ago, when I was a grown man, the owner of a house, two luggers, and as good a diving plant as any man could wish to possess. What was more, just before this I had put some money into a mining concern on the mainland, which had, contrary to most ventures of the sort, turned up trumps, giving me as my share the nice round sum of £5,000. With all this wealth at my back, and having been in harness for a greater number of years on end than I cared to count, I made up my mind to take a holiday and go home to England to see the place where my father was born, and had lived his early life (I found the name of it written in the flyleaf of an old Latin book he left me), and to have a look at a country I'd heard so much about, but never thought to have the good fortune to set my foot upon.

Accordingly I packed my traps, let my house, sold my luggers and gear, intending to buy new ones when I returned, said goodbye to my friends and shipmates, and set off to join an Orient liner in Sydney. You will see from this that I intended doing the thing in style! And why not? I'd got more money to my hand to play with than most of the swells who patronize the first saloon; I had earned it honestly, and was resolved to enjoy myself with it to the top of my bent, and hang the consequences.

I reached Sydney a week before the boat was advertised to sail, but I didn't fret much about that. There's plenty to see and do in such a big place, and when a man's been shut away from theatres and amusements for years at a stretch, he can put in his time pretty well looking about him. All the same, not knowing a soul in the place, I must confess there were moments when I did think regretfully of the tight little island hidden away up north under the wing of New Guinea, of the luggers dancing to the breeze in the harbour, and the warm welcome that always awaited me among my friends in the saloons. Take my word for it, there's something in even being a leader on a small island. Anyway, it's better than being a deadbeat in a big city like Sydney, where nobody knows you, and your next-door neighbour wouldn't miss you if he never saw or heard of you again.

I used to think of these things as I marched about the streets looking in at shop windows, or took excursions up and down the Harbour. There's no place like Sydney Harbour in the wide, wide world for beauty, and before I'd been there a week I was familiar with every part of it. Still, it would have been *more* enjoyable, as I hinted just now, if I had had a friend to tour about with me; and by the same token I'm doing one man an injustice.

There was *one* fellow, I remember, who did offer to show me round: I fell across him in a saloon in George Street. He was tall and handsome, and as spic and span as a new pin till you came to look under

I determine to take a Holiday

the surface. When he entered the bar he winked at the girl who was serving me, and as soon as I'd finished my drink asked me to take another with him. Seeing what his little game was, and wanting to teach him a lesson, I lured him on by consenting. I drank with him, and then he drank with me.

'Been long in Sydney?' he enquired casually, looking at me, and, at the same time, stroking his fair moustache.

'Just come in,' was my reply.

'Don't you find it dull work going about alone?' he enquired. 'I shall never forget my first week of it.'

'You're about right,' I answered. 'It is dull! I don't know a soul, bar my banker and lawyer, in the town.'

'Dear me!' (more curling of the moustache). 'If I can be of any service to you while you're here, I hope you'll command me. For the sake of "Auld Land Syne," dont' you know. I believe we're both Englishmen, eh?'

'It's very good of you,' I replied modestly, affecting to be overcome by his condescension. 'I'm just off to lunch. I am staying at the Quebec. Is it far enough for a hansom?' As he was about to answer, a lawyer, with whom I had done a little business the day before, walked into the room. I turned to my patronizing friend and said, 'Will you excuse me for one moment? I want to speak to this gentleman on business.'

He was still all graciousness.

'I'll call a hansom and wait for you in it.'

When he had left the saloon I spoke to the new arrival. He had noticed the man I had been talking to, and was kind enough to warn me against him.

'That man,' he said, 'bears a very bad reputation. He makes it his trade to meet new arrivals from England—weak-brained young pigeons with money. He shows them round Sydney, and plucks them so clean that, when they leave his hands, in nine cases out of ten, they haven't a feather left to fly with. You ought not, with your experience of rough customers, to be taken in by him.'

'Nor am I,' I replied. 'I am going to teach him a lesson. Would you like to see it? Then come with me.'

Arm in arm we walked into the street, watched by Mr Hawk from his seat in the cab. When we got there we stood for a moment chatting, and then strolled together down the pavement. Next moment I heard the cab coming along after us, and my friend hailing me in his silkiest tones; but though I looked him full in the face I pretended not to know him. Seeing this he drove past us—pulled up a little further down and sprang out to wait for me.

'I was almost afraid I had missed you,' he began, as we came up with him. 'Perhaps as it is such a fine day you would rather walk than ride?'

'I beg your pardon,' I answered; 'I'm really afraid you have the advantage of me.'

'But you have asked me to lunch with you at the Quebec. You told me to call a hansom.'

'Pardon me again! but you are really mistaken. I said I was going to lunch at the Quebec, and asked you if it was far enough to be worth while taking a hansom. That is your hansom, not mine. If you don't require it any longer, I should advise you to pay the man and let him go.'

'You are a swindler, sir. I refuse to pay the cabman. It is your hansom.'

I took a step closer to my fine gentleman, and, looking him full in the face, said as quietly as possible, for I didn't want all the street to hear:

'Mr *Dorunda* Dodson, let this be a lesson to you. Perhaps you'll think twice next time before you try your little games on me!'

He stepped back as if he had been shot, hesitated a moment, and then jumped into his cab and drove off in the opposite direction. When he had gone I looked at my astonished companion.

'Well, now,' he ejaculated at last, 'how on earth did you manage that?'

'Very easily,' I replied. 'I happened to remember having met that gentleman up in our part of the world when he was in a very awkward position—very awkward for him. By his action just now I should say that he has not forgotten the circumstance any more than I have.'

'I should rather think not. Good-day.'

We shook hands and parted, he going on down the street, while I branched off to my hotel.

That was the first of the only two adventures of any importance I met with during my stay in New South Wales. And there's not much in that, I fancy I can hear you saying. Well, that may be so, I don't deny it, but it was nevertheless through that that I became mixed up

I determine to take a Holiday

with the folk who figure in this book, and indeed it was to that very circumstance, and that alone, I owe my connection with the queer story I have set myself to tell. And this is how it came about.

Three days before the steamer sailed, and about four o'clock in the afternoon, I chanced to be walking down Castlereagh Street, wondering what on earth I should do with myself until dinner-time, when I saw approaching me the very man whose discomfiture I have just described. Being probably occupied planning the plucking of some unfortunate new chum, he did not see me. And as I had no desire to meet him again, after what had passed between us, I crossed the road and meandered off in a different direction, eventually finding myself located on a seat in the Domain, lighting a cigarette and looking down over a broad expanse of harbour.

One thought led to another, and so I sat on and on long after dusk had fallen, never stirring until a circumstance occurred on a neighbouring path that attracted my attention. A young and well-dressed lady was pursuing her way in my direction, evidently intending to leave the park by the entrance I had used to come into it. But unfortunately for her, at the junction of two paths to my right, three of Sydney's typical larrikins were engaged in earnest conversation. They had observed the girl coming towards them, and were evidently preparing some plan for accosting her. When she was only about fifty yards away, two of them walked to a distance, leaving the third and biggest ruffian to waylay her. He did so, but without success; she passed him and continued her walk at increased speed.

The man thereupon quickened his pace, and, secure in the knowledge that he was unobserved, again accosted her. Again she tried to escape him, but this time he would not leave her. What was worse, his two friends were now blocking the path in front. She looked to right and left, and was evidently uncertain what to do. Then, seeing escape was hopeless, she stopped, took out her purse, and gave it to the man who had first spoken to her. Thinking this was going too far, I jumped up and went quickly across the turf towards them. My footsteps made no sound on the soft grass, and as they were too much occupied in examining what she had given them, they did not notice my approach.

'You scoundrels!' I said, when I had come up with them. 'What do you mean by stopping this lady? Let her go instantly; and you, my friend, just hand over that purse.'

The man addressed looked at me as if he were taking my measure, and were wondering what sort of chance he'd have against me in a fight. But I suppose my height must have rather scared him, for he changed his tone and began to whine.

'I haven't got the lady's purse, s'help me, I ain't! I was only a asking of 'er the time; I'll take me davy I was!'

'Hand over that purse!' I said sternly, approaching a step nearer to him.

One of the others here intervened—

'Let's stowch 'im, Dog! There ain't a copper in sight!'

With that they began to close upon me. But, as the saying goes, 'I'd been there before.' I'd not been knocking about the rough side of the world for fifteen years without learning how to take care of myself. When they had had about enough of it, which was most likely more than they had bargained for, I took the purse and went down the path to where the innocent cause of it all was standing. She was looking very white and scared, but she plucked up sufficient courage to thank me prettily.

I can see her now, standing there looking into my face with big tears in her pretty blue eyes. She was a girl of about twenty-one or two years of age, I should think—tall, but slenderly built, with a sweet oval face, bright brown hair, and the most beautiful eyes I have ever seen in my life. She was dressed in some dark green material, wore a fawn jacket, and, because the afternoon was cold, had a boa of marten fur round her neck. I can remember also that her hat was of some flimsy make, with lace and glittering spear points in it, and that the whole structure was surmounted by two bows, one of black ribbon, the other of salmon pink.

'Oh, how can I thank you?' she began, when I had come up with her. 'But for your appearance I don't know what those men might not have done to me.'

'I am very glad that I *was* there to help you,' I replied, looking into her face with more admiration for its warm young beauty than perhaps I ought to have shown. 'Here is your purse. I hope you will find its contents safe. At the same time will you let me give you a little piece of advice. From what I have seen this afternoon this is evidently not the sort of place for a young lady to be walking in alone and after dark. I don't think I would risk it again if I were you.'

She looked at me for a moment and then said:

'You are quite right. I have only myself to thank for my misfortune. I met a friend and walked across the green with her; I was on my way back to my carriage—which is waiting for me outside—when I met those men. However, I think I can promise you that it will not happen again, as I am leaving Sydney in a day or two.'

Somehow, when I heard that, I began to feel glad I was booked to leave the place too. But of course I didn't tell her so.

'May I see you safely to your carriage?' I said at last. 'Those fellows may still be hanging about on the chance of overtaking you.'

Her courage must have come back to her, for she looked up into my face with a smile.

'I don't think they will be rude to me again after the lesson you have given them. But if you will walk with me I shall be very grateful.'

Side by side we proceeded down the path, through the gates and out into the street. A neat brougham was drawn up alongside the kerb, and towards this she made her way. I opened the door and held it for her to get in. But before she did so she turned to me and stretched out her little hand.

'Will you tell me your name, that I may know to whom I am indebted?'

'My name is Hatteras. Richard Hatteras, of Thursday Island, Torres Straits. I am staying at the Quebec.'

'Thank you, Mr Hatteras, again and again. I shall always be grateful to you for your gallantry!'

This was attaching too much importance to such a simple action, and I was about to tell her so, when she spoke again:

'I think I ought to let you know who I am. My name is Wetherell, and my father is the Colonial Secretary. I'm sure he will be quite as grateful to you as I am. Goodbye.'

She seemed to forget that we had already shaken hands, for she extended her own a second time. I took it and tried to say something polite, but she stepped into her carriage and shut the door before I could think of anything, and next moment she was being whirled away up the street.

Now old fogies and disappointed spinsters can say what they please about love at first sight. I'm not a romantic sort of person—far from it—the sort of life I had hitherto led was not of a nature calculated to foster a belief in that sort of thing. But if I wasn't over head and ears in love when I resumed my walk that evening, well, I've never known what the passion is.

A daintier, prettier, sweeter little angel surely never walked the earth than the girl I had just been permitted the opportunity of rescuing; and from that moment forward I found my thoughts constantly reverting to her. I seemed to retain the soft pressure of her fingers in mine for hours afterwards, and as a proof of the perturbed state of my feelings I may add that I congratulated myself warmly on having worn that day my new and fashionable Sydney suit, instead of the garments in which I had travelled down from Torres Straits, and which I had hitherto considered quite good enough for even high days and holidays. That she herself would remember me for more than an hour never struck me as being likely.

Next morning I donned my best suit again, gave myself an extra brush up, and sauntered down town to see if I could run across her in the streets. What reason I had for thinking I should is more than I can tell you, but at any rate I was not destined to be disappointed. Crossing George Street a carriage passed me, and in it sat the girl whose fair image had exercised such an effect upon my mind. That she saw and recognized me was evidenced by the gracious bow and smile with which she favoured me. Then she passed out of sight, and it was a wonder that that minute didn't see the end of my career, for I stood like one in a dream looking in the direction in which she had gone, and it was not until two hansoms and a brewer's wagon had nearly run me down that I realized it would be safer for me to pursue my meditations on the sidewalk.

I got back to my hotel by lunch-time, and during the progress of that meal a brilliant idea struck me. Supposing I plucked up courage and called? Why not? It would be only a polite action to enquire if she were any the worse for her fright. The thought was no sooner born in my brain than I was eager to be off. But it was too early for such a formal business, so I had to cool my heels in the hall for an hour. Then, hailing a hansom and enquiring the direction of their residence, I drove off to Potts Point. The house was the last in the street—an imposing mansion standing in well-laid-out grounds. The butler answered my ring, and in response to my enquiry dashed my hopes to the ground by informing me that Miss Wetherell was out.

'She's very busy, you see, at present, sir. She and the master leave for England on Friday in the *Orizaba*.'

'What!' I cried, almost forgetting myself in my astonishment. 'You don't mean to say that Miss Wetherell goes to England in the *Orizaba*?'

'I do, sir. And I do hear she's goin' 'ome to be presented at Court, sir!'

'Ah! Thank you. Will you give her my card, and say that I hope she is none the worse for her fright last evening?'

'He took the card, and a substantial tip with it, and I went back to my cab in the seventh heaven of delight. I was to be shipmates with this lovely creature! For six weeks or more I should be able to see her every day! It seemed almost too good to be true. Instinctively I began to make all sorts of plans and preparations. Who knew but what—but stay, we must bring ourselves up here with a round turn, or we shall be anticipating what's to come.

To make a long story short—for it must be remembered that what I am telling you is only the prelude to all the extraordinary things that will have to be told later on—the day of sailing came. I went down to the boat on the morning of her departure, and got my baggage safely stowed away in my cabin before the rush set in. My cabin mate was to join the ship in Adelaide, so for the first few days of the voyage I should be alone.

About three o'clock we hove our anchor and steamed slowly down the Bay. It was a perfect afternoon, and the Harbour, with its multitudinous craft of all nationalities and sizes, the blue water backed by stately hills, presented a scene the beauty of which would have appealed to the mind of the most prosaic. I had been below when the Wetherells arrived on board, so the young lady had not yet become aware of my presence. Whether she would betray any astonishment when she did find out was beyond my power to tell; at any rate, I know that I was by a long way the happiest man aboard the boat that day. However, I was not to be kept long in suspense. Before we had reached the Heads it was all settled, and satisfactorily so. I was standing on the promenade deck, just abaft the main saloon entrance, watching the panorama spread out before me, when I heard a voice I recognized only too well say behind me:

'And so goodbye to you, dear old Sydney. Great things will have happened when I set eyes on you again.'

Little did she know how prophetic were her words. As she spoke I turned and confronted her. For a moment she was overwhelmed with surprise, then, stretching out her hand, she said:

'Really, Mr Hatteras, this is most wonderful. You are the last person I expected to meet on board the *Orizaba*.'

'And perhaps,' I replied, 'I might with justice say the same of you. It looks as if we are destined to be fellow-travellers.'

She turned to a tall, white-bearded man beside her.

'Papa, I must introduce you to Mr Hatteras. You will remember I told you how kind Mr Hatteras was when those larrikins were rude to me in the Domain.'

'I am sincerely obliged to you, Mr Hatteras,' he said, holding out his hand and shaking mine heartily. 'My daughter did tell me, and I called yesterday at your hotel to thank you personally, but you were unfortunately not at home. Are you visiting Europe?'

'Yes; I'm going home for a short visit to see the place where my father was born.'

'Are you then, like myself, an Australian native? I mean, of course, as you know, colonial born?' asked Miss Wetherell with a little laugh. The idea of her calling herself an Australian native in any other sense! The very notion seemed preposterous.

'I was born at sea, a degree and a half south of Mauritius,' I answered; 'so I don't exactly know what you would call me. I hope you have comfortable cabins?'

'Very. We have made two or three voyages in this boat before, and we always take the same places. And now, papa, we must really go and see where poor Miss Thompson is. We are beginning to feel the swell, and she'll be wanting to go below. Goodbye for the present, Mr Hatteras.'

I raised my cap and watched her walk away down the deck, balancing herself as if she had been accustomed to a heaving plank all her life. Then I turned to watch the fast receding shore, and to my own thoughts, which were none of the saddest, I can assure you. For it must be confessed here, and why should I deny it? that I was in love from the soles of my deck shoes to the cap upon my head. But as to the chance that I, a humble pearler, would stand with one of Sydney's wealthiest and most beautiful daughters—why, that's another matter, and one that, for the present, I was anxious to keep behind me.

I determine to take a Holiday 23

Within the week we had left Adelaide behind us, and four days later Albany was also a thing of the past. By the time we had cleared the Lewin we had all settled down to our life aboard ship, the bad sailors were beginning to appear on deck again, and the medium voyagers to make various excuses for their absences from meals. One thing was evident, that Miss Wetherell was the belle of the ship. Everybody paid her attention, from the skipper down to the humblest deck hand. And this being so, I prudently kept out of the way, for I had no desire to be thought to presume on our previous acquaintance. Whether she noticed this I cannot tell, but at any rate her manner to me when we *did* speak was more cordial than I had any right or reason to expect it would be. Seeing this, there were not wanting people on board who scoffed and sneered at the idea of the Colonial Secretary's daughter noticing so humble a person as myself, and when it became known what my exact social position was, I promise you these malicious whisperings did not cease.

One evening, two or three days after we had left Colombo behind us, I was standing at the rails on the promenade deck a little abaft the smoking-room entrance, when Miss Wetherell came up and took her place beside me. She looked very dainty and sweet in her evening dress, and I felt, if I had known her better, I should have liked to tell her so.

'Mr Hatteras,' said she, when we had discussed the weather and the sunset, 'I have been thinking lately that you desire to avoid me.'

'Heaven forbid! Miss Wetherell,' I hastened to reply. 'What on earth can have put such a notion into your head?'

'All the same, I believe it to be true. Now, why do you do it?'

'I have not admitted that I do it. But, perhaps, if I *do* seem to deny myself the pleasure of being with you as much as some other people I could mention, it is only because I fail to see what possible enjoyment you can derive from my society.'

'That is a very pretty speech,' she answered, smiling, 'but it does not tell me what I want to know.'

'And what is it that you want to know, my dear young lady?'

'I want to know why you are so much changed towards me. At first we got on splendidly—you used to tell me of your life in Torres Straits, of your trading ventures in the Southern Seas, and even of your hopes for the future. Now, however, all that is changed. It is

"Good-morning, Miss Wetherell," "Good-evening, Miss Wetherell," and that is all. I must own I don't like such treatment.'

'I must crave your pardon—but——'

'No, we won't have any "buts". If you want to be forgiven, you must come and talk to me as you used to do. You will like the rest of the people I'm sure when you get to know them. They are very kind to me.'

'And you think I shall like them for that reason?'

'No, no. How silly you are! But I do so want you to be friendly.'

After that there was nothing for it but for me to push myself into a circle where I had the best reasons for knowing that I was not wanted. However, it had its good side: I saw more of Miss Wetherell; so much more indeed that I began to notice that her father did not quite approve of it. But, whatever he may have thought, he said nothing to me on the subject.

A fortnight or so later we were at Aden, leaving that barren rock about four o'clock, and entering the Red Sea the same evening. The Suez Canal passed through, and Port Said behind us, we were in the Mediterranean, and for the first time in my life I stood in Europe.

At Naples the Wetherells were to say goodbye to the boat, and continue the rest of their journey home across the Continent. As the hour of separation approached, I must confess I began to dread it more and more. And somehow, I fancy, *she* was not quite as happy as she used to be. You will probably ask what grounds I had for believing that a girl like Miss Wetherell would take any interest in a man like myself; and it is a question I can no more answer than I can fly. And yet, when I came to think it all out, I was not without my hopes.

We were to reach port the following morning. The night was very still, the water almost unruffled. Somehow it came about that Miss Wetherell and I found ourselves together in the same sheltered spot where she had spoken to me on the occasion referred to before. The stars in the east were paling preparatory to the rising of the moon. I glanced at my companion as she leant against the rails scanning the quiet sea, and noticed the sweet wistfulness of her expression. Then, suddenly, a great desire came over me to tell her of my love. Surely, even if she could not return it, there would be no harm in letting her know how I felt towards her. For this reason I drew a little closer to her.

I determine to take a Holiday

'And so, Miss Wetherell,' I said, 'tomorrow we are to bid each other goodbye; never, perhaps, to meet again.'

'Oh, no, Mr Hatteras,' she answered, 'we won't say that. Surely we shall see something of each other somewhere. The world is very tiny after all.'

'To those who desire to avoid each other, perhaps, but for those who wish to *find* it is still too large.'

'Well, then, we must hope for the best. Who knows but that we may run across each other in London. I think it is very probable.'

'And will that meeting be altogether distasteful to you?' I asked, quite expecting that she would answer with her usual frankness. But to my surprise she did not speak, only turned half away from me. Had I offended her?

'Miss Wetherell, pray forgive my rudeness,' I said hastily. 'I ought to have known I had no right to ask you such a question.'

'And why shouldn't you?' she replied, this time turning her sweet face towards me. 'No, Mr Hatteras, I will tell you frankly, I should very much like to see you again.'

With that all the blood in my body seemed to rush to my head. Could I be dreaming? Or had she really said she would like to see me again? I would try my luck now whatever came of it.

'You cannot think how pleasant our intercourse has been to me,' I said. 'And now I have to go back to my lonely, miserable existence again.'

'But you should not say that; you have your work in life!'

'Yes, but what is that to me when I have no one to work for? Can you conceive anything more awful than my loneliness? Remember as far as I know I am absolutely without kith and kin. There is not a single soul to care for me in the whole world—not one to whom my death would be a matter of the least concern.'

'Oh, don't—don't say that!'

Her voice faltered so that I turned from the sea and contemplated her.

'It is true, Miss Wetherell, bitterly true.'

'It is not true. It cannot be true!'

'If only I could think it would be some little matter of concern to you I should go back to my work with a happier heart.'

Again she turned her face from me. My arm lay beside hers upon the bulwarks, and I could feel that she was trembling. Brutal though it may seem to say so, this gave me fresh courage. I said slowly, bending my face a little towards her:

'Would it affect you, Phyllis?'

One little hand fell from the bulwarks to her side, and as I spoke I took possession of it. She did not appear to have heard my question, so I repeated it. Then her head went down upon the bulwarks, but not before I had caught the whispered 'yes' that escaped her lips.

Before she could guess what was going to happen, I had taken her in my arms and smothered her face with kisses. Nor did she offer any resistance. I knew the whole truth now. She was mine, she loved me—me—me—me! The whole world seemed to re-echo the news, the very sea to ring with it, and just as I learned from her own dear lips the story of her love, the great moon rose as if to listen. Can you imagine my happiness, my delight? She was mine, this lovely girl, my very own! bound to me by all the bonds of love. Oh, happy hour! Oh, sweet delight!

I pressed her to my heart again and again. She looked into my face and then away from me, her sweet eyes suffused with tears, then suddenly her expression changed. I turned to see what ailed her, and to my discomfiture discovered her father stalking along the silent deck towards us.

Whispering to her to leave us she sped away, and I was left alone with her angry parent. That he *was* angry I judged from his face; nor was I wrong in my conjecture.

'Mr Hatteras,' he said severely, 'pray what does this mean? How is it that I find you in this undignified position with my daughter?'

'Mr Wetherell,' I answered, 'I can see that an explanation is due to you. Just before you came up I was courageous enough to tell your daughter that I loved her. She has been generous enough to inform me that she returns my affection. And now the best course for me to pursue is to ask your permission to make her my wife.'

'You presume, sir, upon the service you rendered my daughter in Sydney. I did not think you would follow it up in this fashion.'

'Your daughter is free to love whom she pleases, I take it,' I said, my temper, fanned by the tone he adopted, getting a little the better of my judgement. 'She has been good enough to promise to marry me—if I can obtain your permission. Have you any objection to raise?'

'Only one, and that one is insuperable! Understand me, I forbid it once and for all! In every particular—without hope of change—I forbid it!'

'As you must see it is a matter which affects the happiness of two lives, I feel sure you will be good enough to tell me your reasons?'

'I must decline any discussion on the matter at all. You have my answer, I forbid it!'

'This is to be final, then? I am to understand that you are not to be brought to change your mind by any actions of mine?'

'No, sir, I am not! What I have said is irrevocable. The idea is not to be thought of for a moment. And while I am on this subject let me tell you that your conduct towards my daughter on board this ship has been very distasteful to me. I have the honour to wish you a very good-evening.'

'Stay, Mr Wetherell,' I said, as he turned to go. 'You have been kind enough to favour me with your views. Now I will give you mine. Your daughter loves me. I am an honest and an industrious man, and I love her with my whole heart and soul. I tell you now, and though you decline to treat me with proper fairness, I give you warning that I intend to marry her if she will still have me—with your consent or without it!'

'You are insolent, sir.'

'I assure you I have no desire to be. I endeavour to remember that you are her father, though I must own you lack her sense of what is fair and right.'

'I will not discuss the question any further with you. You know my absolute decision. Good-night!'

'Good-night!'

With anger and happiness struggling in my breast for the mastery, I paced that deck for hours. My heart swelled with joy at the knowledge that my darling loved me, but it sank like lead when I considered the difficulties which threatened us if her father persisted in his present determination. At last, just as eight bells was striking (twelve o'clock), I went below to my cabin. My fellow-passenger was fast asleep—a fact which I was grateful for when I discovered propped against my bottle-rack a tiny envelope with my name inscribed upon it. Tearing it open I read the following:

My own Dearest,

My father has just informed me of his interview with you. I cannot understand it or ascribe a reason for it. But whatever happens, remember that I will be your wife, and the wife of no other.

May God bless and keep you always.

Your own,
Phyllis.

P.S.—Before we leave the ship you must let me know your address in London.

With such a letter under my pillow, can it be doubted that my dreams were good? How little I guessed the accumulation of troubles to which this little unpleasantness with Mr Wetherell was destined to be the prelude!

CHAPTER II

London

Now that I come to think the matter out, I don't know that I could give you any definite idea of what my first impressions of London were. One thing at least is certain, I had never had experience of anything approaching such a city before, and, between ourselves, I can't say that I ever want to again. The constant rush and roar of traffic, the crowds of people jostling each other on the pavements, the happiness and the misery, the riches and the poverty, all mixed up together in one jumble, like good and bad fruit in a basket, fairly took my breath away; and when I went down, that first afternoon, and saw the Park in all its summer glory, my amazement may be better imagined than described.

I could have watched the carriages, horsemen, and promenaders for hours on end without any sense of weariness. And when a bystander, seeing that I was a stranger, took compassion upon my ignorance and condescended to point out to me the various celebrities present, my pleasure was complete. There certainly is no place like London for show and glitter, I'll grant you that; but all the same I'd no more think of taking up my permanent abode in it than I'd try to cross the Atlantic in a Chinese sampan.

Having before I left Sydney been recommended to a quiet hotel in a neighbourhood near the Strand, convenient both for sightseeing and business, I had my luggage conveyed thither, and prepared to make myself comfortable for a time. Every day I waited eagerly for a letter from my sweetheart, the more impatiently because its non-arrival convinced me that they had not yet arrived in London. As it turned out, they had delayed their departure from Naples for two days, and had spent another three in Florence, two in Rome, and a day and a half in Paris.

One morning, however, my faithful watch over the letter rack, which was already becoming a standing joke in the hotel, was rewarded. An envelope bearing an English stamp and postmark, and addressed in a handwriting as familiar to me as my own, stared me in the face. To take it out and break the seal was the work of a moment. It was only a matter of a few lines, but it brought me news that raised me to the seventh heaven of delight.

Mr and Miss Wetherell had arrived in London the previous afternoon, they were staying at the Hôtel Métropole, would leave town for the country at the end of the week, but in the meantime, if I wished to see her, my sweetheart would be in the entrance hall of the British Museum the following morning at eleven o'clock.

How I conducted myself in the interval between my receipt of the letter and the time of the appointment, I have not the least remembrance; I know, however, that half-past ten, on the following morning, found me pacing up and down the street before that venerable pile, scanning with eager eyes every conveyance that approached me from the right or left. The minutes dragged by with intolerable slowness, but at length the time arrived.

A kindly church clock in the neighbourhood struck the hour, and others all round it immediately took up the tale. Before the last stroke had died away a hansom turned towards the gates from Bury Street, and in it, looking the picture of health and dainty beauty, sat the girl who, I had good reason to know, was more than all the world to me. To attract her attention and signal to the driver to pull up was the work of a second, and a minute later I had helped her to alight, and we were strolling together across the square towards the building.

'Ah, Dick,' she said, with a roguish smile, in answer to a question of mine, 'you don't know what trouble I had to get away this morning. Papa had a dozen places he wished me to go to with him. But when I told him that I had some very important business of my own to attend to before I could go calling, he was kind enough to let me off.'

'I'll be bound he thought you meant business with a dressmaker,' I laughingly replied, determined to show her that I was not unversed in the ways of women.

'I'm afraid he did,' she answered, blushing, 'and for that very reason alone I feel horribly guilty. But my heart told me I must see you at once, whatever happened.'

Could any man desire a prettier speech than that? If so, I was not that man. We were inside the building by this time, ascending the great staircase. A number of pretty, well-dressed girls were to be seen moving about the rooms and corridors, but not one who could in any way compare with the fair Australian by my side.

As we entered the room at the top of the stairs, I thought it a good opportunity to ask the question I had been longing to put to her.

'Phyllis, my sweetheart,' I said, with almost a tremor in my voice, 'it is a fortnight now since I spoke to you. You have had plenty of time to consider our position. Have you regretted giving me your love?'

We came to a standstill, and leant over a case together, but what it contained I'm sure I haven't the very vaguest idea.

She looked up into my face with a sweet smile.

'Not for one single instant, Dick! Having once given you my love, is it likely I should want it back again?'

'I don't know. Somehow I can't discover sufficient reason for your giving it to me at all.'

'Well, be sure I'm not going to tell you. You might grow conceited. Isn't it sufficient that I *do* love you, and that I am not going to give you up, whatever happens?'

'More than sufficient,' I answered solemnly. 'But, Phyllis, don't you think I can induce your father to relent? Surely as a good parent he must be anxious to promote your happiness at any cost to himself?'

'I can't understand it at all. He has been so devoted to me all my life that his conduct now is quite inexplicable. Never once has he denied me anything I really set my heart upon, and he always promised me that I should be allowed to marry whomsoever I pleased, provided he was a good and honourable man, and one of whom he could in any way approve. And you are all that, Dick, or I shouldn't have loved you, I know.'

'I don't think I'm any worse than the ordinary run of men, dearest, if I am no better. At any rate I love you with a true and honourable love. But don't you think he will come round in time?'

'I'm almost afraid not. He referred to it only yesterday, and seemed quite angry that I should have dared to entertain any thought of you after what he said to me on board ship. It was the first time in my life he ever spoke to me in such a tone, and I felt it keenly. No, Dick, there is something behind it all that I cannot understand. Some mystery that I would give anything to fathom. Papa has not been himself ever since we started for England. Indeed, his very reason for coming at all is an enigma to me. And now that he *is* here, he seems in continual dread of meeting somebody—but who that somebody is, and why my father, who has the name and reputation of being such a courageous, determined, honourable man, should be afraid, is a thing I cannot understand.'

'It's all very mysterious and unfortunate. But surely something can be done? Don't you think if I were to see him again, and put the matter more plainly before him, something might be arranged?'

'It would be worse than useless at present, I fear. No, you must just leave it to me, and I'll do my best to talk him round. Ever since my mother died I have been as his right hand, and it will be strange if he does not listen to me and see reason in the end.'

Seeing who it was that would plead with him I did not doubt it.

By this time we had wandered through many rooms, and now found ourselves in the Egyptian Department, surrounded by embalmed dead folk and queer objects of all sorts and descriptions. There was something almost startling about our love-making in such a place, among these men and women, whose wooings had been conducted in a country so widely different to ours, and in an age that was dead and gone over two thousand years ere we were born. I spoke of this to Phyllis. She laughed and gave a little shiver.

'I wonder,' she said, looking down on the swathed-up figure of a princess of the royal house of Egypt, lying stretched out in the case beside which we sat, 'if this great lady, who lies so still and silent now, had any trouble with her love affair?'

'Perhaps she had more than one beau to her string, and not being allowed to have one took the other,' I answered; 'though from what we can see of her now she doesn't look as if she were ever capable of exercising much fascination, does she?'

As I spoke I looked from the case to the girl and compared the swaddled-up figure with the healthy, living, lovely creature by my side. But I hadn't much time for comparison. My sweetheart had taken her watch from her pocket and was glancing at the dial.

'A quarter to twelve!' she cried in alarm. 'Oh, Dick, I must be going. I promised to meet papa at twelve, and whatever happens I must not keep him waiting.'

She rose and was about to pull on her gloves. But before she had time to do so I had taken a little case from my pocket and opened it. When she saw what it contained she could not help a little womanly cry of delight.

'Oh, Dick! you naughty, extravagant boy!'

'Why, dearest? Why naughty or extravagant to give the woman I love a little token of my affection?' As I spoke I slipped the ring over her pretty finger and raised the hand to my lips.

'Will you try,' I said, 'whenever you look at that ring, to remember that the man who gave it to you loves you with his whole heart and soul, and will count no trouble too great, or no exertion too hard, to make you happy?'

'I will remember,' she said solemnly, and when I looked I saw that tears stood in her eyes. She brushed them hastily away, and after an interlude which it hardly becomes me to mention here, we went down the stairs again and out into the street, almost in silence.

Having called a cab, I placed her in it and nervously asked the question that had been sometime upon my mind:

'When shall I see you again?'

'I cannot tell,' she answered. 'Perhaps next week. But I'll let you know. In the meantime don't despair; all will come right yet! Goodbye.'

'Goodbye and God bless you!'

I lifted my hat, she waved her hand, and next moment the hansom had disappeared round the corner.

Having seen the last of her I wandered slowly down the pavement towards Oxford Street, then turning to my left hand, made my way citywards. My mind was full of my interview with the sweet girl who had just left me, and I wandered on and on, wrapped in my own thoughts, until I found myself in a quarter of London into which I had never hitherto penetrated. The streets were narrow, and, as if to be in keeping with the general air of gloom, the shops were small and their wares of a peculiarly sordid nature; hand-carts, barrows, and stalls lined the grimy pavements, and the noise was deafening.

A church clock somewhere in the neighbourhood struck 'One', and as I was beginning to feel hungry, and knew myself to be a long way from my hotel, I cast about me for a lunching-place. But it was some time before I encountered the class of restaurant I wanted. When I did it was situated at the corner of two streets, carried a foreign name over the door, and, though considerably the worse for wear, presented a cleaner appearance than any other I had as yet experienced.

Pushing the door open I entered. An unmistakable Frenchman, whose appearance, however, betokened long residence in England, stood behind a narrow counter polishing an absinthe glass. He bowed politely and asked my business.

'Can I have lunch?' I asked.

'Oui, monsieur! Cer-tain-lee. If monsieur will walk upstairs I will take his order.'

Waving his hand in the direction of a staircase in the corner of the shop he again bowed elaborately, while I, following the direction he indicated, proceeded to the room above. It was long and lofty, commanded an excellent view of both thoroughfares, and was furnished with a few inferior pictures, a much worn oilcloth, half a dozen small marble-top tables, and four times as many chairs.

When I entered three men were in occupation. Two were playing chess at a side table, while the third, who had evidently no connection with them, was watching the game from a distance, at the same time pretending to be absorbed in his paper. Seating myself at a table near the door, I examined the bill of fare, selected my lunch, and in order to amuse myself while it was preparing, fell to scrutinizing my companions.

Of the chess-players, one was a big, burly fellow, with enormous arms, protruding rheumy eyes, a florid complexion, and a voluminous red beard. His opponent was of a much smaller build, with pale features, a tiny moustache, and watery blue eyes. He wore a pince-nez, and from the length of his hair and a dab of crimson lake upon his shirt cuff, I argued him an artist.

Leaving the chess-players, my eyes lighted on the stranger on the other side. He was more interesting in every way. Indeed, I was surprised to see a man of his stamp in the house at all. He was tall and slim, but exquisitely formed, and plainly the possessor of enormous strength. His head, if only from a phrenological point of view, was a magnificent one, crowned with a wealth of jet black hair. His eyes were dark as night, and glittered like those of a snake. His complexion was of a decidedly olive hue, though, as he sat in the shadow of the corner, it was difficult to tell this at first sight.

But what most fascinated me about this curious individual was the interest he was taking in the game the other men were playing. He kept his eyes fixed upon the board, looked anxiously from one to the other as a move trembled in the balance, smiled sardonically when his desires were realized, and sighed almost aloud when a mistake was made.

Every moment I expected his anxiety or disappointment to find vent in words, but he always managed to control himself in time. When he became excited I noticed that his whole body quivered under its influence, and once when the smaller of the players made an injudicious move a look flew into his face that was full of such malignant intensity that I'll own I was influenced by it. What effect it would have had upon the innocent cause of it all, had he seen it, I should have been sorry to conjecture.

Just as my lunch made its appearance the game reached a conclusion, and the taller of the two players, having made a remark in German, rose to leave. It was evident that the smaller man had won, and in an excess of pride, to which I gathered his nature was not altogether a stranger, he looked round the room as if in defiance.

Doing so, his eyes met those of the man in the corner. I glanced from one to the other, but my gaze rested longest on the face of the smaller man. So fascinated did he seem to be by the other's stare that his eyes became set and stony. It was just as if he were being mesmerized. The person he looked at rose, approached him, sat down at the table and began to arrange the men on the board without a word. Then he looked up again.

'May I have the pleasure of giving you a game?' he asked in excellent English, bowing slightly as he spoke, and moving a pawn with his long white fingers.

The little man found voice enough to murmur an appropriate reply, and they began their game, while I turned to my lunch. But, in spite of myself, I found my eyes continually reverting to what was happening at the other table. And, indeed, it was a curious sight I saw there.

The tall man had thrown himself into the business of the game, heart and soul. He half sat, half crouched over the board, reminding me more of a hawk hovering over a poultry yard than anything else I can liken him to.

His eyes were riveted first on the men before him and then on his opponent—his long fingers twitched and twined over each move, and seemed as if they would never release their hold. Not once did he speak, but his attitude was more expressive than any words.

The effect on the little man, his companion, was overwhelming. He was quite unable to do anything, but sat huddled up in his chair as if terrified by his demoniacal companion. The result even a child might have foreseen. The tall man won, and the little man, only too glad to have come out of the ordeal with a whole skin, seized his hat and, with a half-uttered apology, darted from the room.

For a moment or two his extraordinary opponent sat playing with the chessmen. Then he looked across at me and without hesitation said, accompanying his remark with a curious smile, for which I could not at all account:

'I think you will agree with me that the limitations of the fool are the birth gifts of the wise!'

Not knowing what reply to make to this singular assertion, I wisely held my tongue. This brought about a change in his demeanour; he rose from his seat, and came across to where I sat. Seating himself in a chair directly opposite me, he folded his hands in his lap, after the manner of a demure old spinster, and, having looked at me earnestly, said with an almost indescribable sweetness of tone:

'I think you will allow, Mr Hatteras, that half the world is born for the other half to prey upon!'

For a moment I was too much astonished to speak; how on earth had he become aware of my name? I stumbled out some sort of reply, which evidently did not impress him very much, for he began again:

'Our friend who has just left us will most certainly be one of those preyed upon. I pity him because he will not have the smallest grain of pleasure in his life. You, Mr Hatteras, on the other hand, will, unwittingly, be in the other camp. Circumstances will arrange that for you. Some have, of course, no desire to prey; but necessity forces it on them. Yourself, for instance. Some only prey when they are quite sure there will be no manner of risk. Our German friend who played the previous game is an example. Others, again, never lose an opportunity. Candidly speaking, to which class should you imagine I belong?'

He smiled as he put the question, and, his thin lips parting, I could just catch the glitter of the short teeth with which his mouth was furnished. For the third time since I had made his acquaintance I did not know which way to answer. However, I made a shot and said something.

'I really know nothing about you,' I answered. 'But from your kindness in giving our artist friend a game, and now in allowing me the benefit of your conversation, I should say you only prey upon your fellow-men when dire extremity drives you to it.'

'And you would be wrong. I am of the last class I mentioned. There is only one sport of any interest to me in life, and that is the opportunity of making capital out of my fellow humans. You see, I am candid with you, Mr Hatteras!'

'Pray excuse me. But you know my name! As I have never, to my knowledge, set eyes on you before, would you mind telling me how you became acquainted with it?'

'With every pleasure. But before I do so I think it only fair to tell you that you will not believe my explanation. And yet it *should* convince you. At any rate we'll try. In your right-hand top waistcoat pocket you have three cards.' Here he leant his head on his hands and shut his eyes. 'One is crinkled and torn, but it has written on it, in pencil, the name of Edward Braithwaite, Macquarrie Street, Sydney. I presume the name is Braithwaite, but the *t* and *e* are almost illegible. The second is rather a high sounding one—the Hon. Sylvester Wetherell, Potts Point, Sydney, New South Wales, and the third is, I take it, your own, Richard Hatteras. Am I right?'

I put my fingers in my pocket, and drew out what it contained—a half-sovereign, a shilling, a small piece of pencil, and three cards. The first, a well-worn piece of pasteboard, bore, surely enough, the name of Edward Braithwaite, and was that of the solicitor with whom I transacted my business in Sydney; the second was given me by my sweetheart's father the day before we left Australia; and the third was certainly my own.

Was this witchcraft or only some clever conjuring trick? I asked myself the question, but could give it no satisfactory answer. At any rate you may be sure it did not lessen my respect for my singular companion.

'Ah! I am right then!' he cried exultingly. 'Isn't it strange how the love of being right remains with us, when we think we have safely combated every other self-conceit. Well, Mr Hatteras, I am very pleased to have made your acquaintance. Somehow I think we are destined to meet again—where I cannot say. At any rate, let us hope that that meeting will be as pleasant and successful as this has been.'

But I hardly heard what he said. I was still puzzling my brains over his extraordinary conjuring trick—for trick I am convinced it was. He had risen and was slowly drawing on his gloves when I spoke.

'I have been thinking over those cards,' I said, 'and I am considerably puzzled. How on earth did you know they were there?'

'If I told you, you would have no more faith in my powers. So with your permission I will assume the virtue of modesty. Call it a conjuring trick, if you like. Many curious things are hidden under that comprehensive term. But that is neither here nor there. Before I go would you like to see one more?'

'Very much, indeed, if it's as good as the last!' I replied.

In the window stood a large glass dish, half full of water and having a dark brown fly paper floating on the surface. He brought it across to the table at which I sat, and having drained the water into a jug near by, left the paper sticking to the bottom.

This done, he took a tiny leather case from his pocket and a small bottle out of that again. From this bottle he poured a few drops of some highly pungent liquid on to the paper, with the result that it grew black as ink and threw off a tiny vapour, which licked the edges of the bowl and curled upwards in a faint spiral column.

'There, Mr Hatteras, this is a—well, a trick—I learned from an old woman in Benares. It is a better one than the last and will repay your interest. If you will look on that paper for a moment, and try to concentrate your attention, you will see something that will, I think, astonish you.'

Hardly believing that I should see anything at all I looked. But for some seconds without success. My scepticism, however, soon left me. At first I saw only the coarse grain of the paper and the thin vapour rising from it. Then the knowledge that I was gazing into a dish vanished. I forgot my companion and the previous conjuring trick. I saw only a picture opening out before me—that of a handsomely furnished room, in which was a girl sitting in an easy chair crying as if her heart were breaking. The room I had never seen before, but the girl I should have known among a thousand. *She was Phyllis, my sweetheart!*

I looked and looked, and as I gazed at her I heard her call my name. 'Oh, Dick! Dick! come to me!' Instantly I sprang to my feet, meaning to cross the room to her. Next moment I became aware of a loud crash. The scene vanished, my senses came back to me, and to my astonishment I found myself standing alongside the overturned restaurant table. The glass dish lay on the floor shattered into a thousand fragments. My friend, the conjuror, had disappeared.

Having righted the table again, I went downstairs and explained my misfortune. When I had paid my bill I took my departure, more troubled in mind than I cared to confess. That it was only what he had called it, a conjuring trick, I felt I ought to be certain, but still it was clever and uncanny enough to render me very uncomfortable.

In vain I tried to drive the remembrance of the scene I had witnessed from my brain, but it would not be dispelled. At length, to satisfy myself, I resolved that if the memory of it remained with me so vividly in the morning I would take the bull by the horns and call at the Métropole to make enquiries.

I returned to my hotel in time for dinner, but still I could not rid myself of the feeling that some calamity was approaching. Having sent my meal away almost untouched, I called a hansom and drove to the nearest theatre, but the picture of Phyllis crying and calling for me in vain kept me company throughout the performance, and brought me home more miserable at the end than I had started. All night long I dreamed of it, seeing the same picture again and again, and hearing the same despairing cry, 'Oh, Dick! Dick! come to me!'

In the morning there was only one thing to be done. Accordingly, after breakfast I set off to make sure that nothing was the matter. On the way I tried to reason with myself. I asked how it was that I, Dick Hatteras, a man who thought he knew the world so well, should have been so impressed with a bit of wizardry as to be willing to risk making a fool of myself before the two last people in the world I wanted to think me one. Once I almost determined to turn back, but while the intention held me the picture rose again before my mind's eye, and on I went more resolved to solve the mystery than before.

Arriving at the hotel, I paid my cabman and entered the hall. A gorgeously caparisoned porter stood on the steps, and of him I enquired where I could find Miss Wetherell. Imagine my surprise when he replied:

'*They've left, sir. Started yesterday afternoon, quite suddenly, for Paris, on their way back to Australia!*'

CHAPTER III

I Visit my Relations

For the moment I could hardly believe my ears. Gone? Why had they gone? What could have induced them to leave England so suddenly? I questioned the hall porter on the subject, but he could tell me nothing save that they had departed for Paris the previous day, intending to proceed across the Continent in order to catch the first Australian boat at Naples.

Feeling that I should only look ridiculous if I stayed questioning the man any longer, I pressed a tip into his hand and went slowly back to my own hotel to try and think it all out. But though I devoted some hours to it, I could arrive at no satisfactory conclusion. The one vital point remained and was not to be disputed—they were gone. But the mail that evening brought me enlightenment in the shape of a letter, written in London and posted in Dover. It ran as follows:

Monday Afternoon.

MY OWN DEAREST.—Something terrible has happened to papa! I cannot tell you what, because I do not know myself. He went out this morning in the best of health and spirits, and returned half an hour ago trembling like a leaf and white as a sheet. He had only strength enough left to reach a chair in my sitting-room before he fainted dead away. When he came to himself again he said, 'Tell your maid to pack at once. There is not a moment to lose. We start for Paris this evening to catch the next boat leaving Naples for Australia.' I said, 'But, papa!' 'Not a word,' he answered. 'I have seen somebody this morning whose presence renders it impossible for us to remain an instant longer in England. Go and pack at once, unless you wish my death to lie at your door.' After that I could, of course, say nothing. I have packed, and now, in half an hour, we leave England again. If I could only see you to say goodbye; but that, too, is impossible. I cannot tell what it all means, but that it is very serious business that takes us away so suddenly I feel convinced. My father seems frightened to remain in London a minute longer than he can help. He even stands at the window as I write, earnestly scrutinizing everybody who enters the hotel. And now, my own——

But what follows, the reiterations of her affection, her vows to be true to me, etc., etc., could have no possible interest for any one save lovers. And even those sympathetic ones I have, unfortunately, not the leisure now to gratify.

I sat like one stunned. All enjoyment seemed suddenly to have gone out of life for me. I could only sit twirling the paper in my hand and picturing the train flying remorselessly across France, bearing away from me the girl I loved better than all the world. I went down to the Park, but the scene there had no longer any interest in my eyes. I went later on to a theatre, but I found no enjoyment in the piece performed. London had suddenly become distasteful to me. I felt I must get out of it; but where could I go? Every place was alike in my present humour. Then one of the original motives of my journey rose before me, and I determined to act on the suggestion.

Next morning I accordingly set off for Hampshire to try, if possible, to find my father's old home. What sort of a place it would turn out to be I had not the very remotest idea. But I'd got the address by heart, and, with the help of a Bradshaw, for that place I steered.

Leaving the train at Lyndhurst Road—for the village I was in search of was situated in the heart of the New Forest—I hired a ramshackle conveyance from the nearest innkeeper and started off for it. The man who drove me had lived in the neighbourhood, so he found early occasion to inform me, all his seventy odd years, and it struck him as a humorous circumstance that he had never in his life been even as far as Southampton, a matter of only a few miles by road and ten minutes by rail.

And that self-same sticking at home is one of the things about England and yokel Englishmen that for the life of me I cannot understand. It seems to me—of course, I don't put it forward that I'm right—that a man might just as well be dead as only know God's world for twenty miles around him. It argues a poverty of interest in

I Visit my Relations

the rest of creation—a sort of mud-turtle existence, that's neither encouraging nor particularly ornamental. And yet if everybody went a-travelling where would the prosperity of England be? That's a point against my argument, I must confess. Well, perhaps we had travelled a matter of two miles when it struck me to ask my charioteer about the place to which we were proceeding. It was within the bounds of possibility, I thought, that he might once have known my father. I determined to try him. So waiting till we had passed a load of hay coming along the lane, I put the question to him.

To my surprise, he had no sooner heard the name than he became as excited as it was possible for him to be.

'Hatteras!' he cried. 'Be ye a Hatteras? Well, well, now, dearie me, who'd ha' thought it!'

'Do you know the name so well, then?'

'Ay! ay! I know the name well enough; who doesn't in these parts? There was the old squire and Lady Margaret when first I remember. Then Squire Jasper and his son, the captain, as was killed in the mutiny in foreign parts—and Master James——'

'James—that was my father's name. James Dymoke Hatteras.'

'You Master James' son—you don't say! Well! well! Now to think of that too! Him that ran away from home after words with the Squire and went to foreign parts. Who'd have thought it! Lawksee me! Sir William will be right down glad to see ye, I'll be bound.'

'Sir William, and who's Sir William?'

'He's the only one left now, sir. Lives up at the House. Ah, dear! ah, dear! There's been a power o' trouble in the family these years past.'

By this time the aspect of the country was changing. We had left the lane behind us, ascended a short hill, and were now descending it again through what looked to my eyes more like a stately private avenue than a public road. Beautiful elms reared themselves on either hand and intermingled their branches overhead; while before us, through a gap in the foliage, we could just distinguish the winding river, with the thatched roofs of the village, of which we had come in search, lining its banks, and the old grey tower of the church keeping watch and ward over all.

There was to my mind something indescribably peaceful and even sad about that view, a mute sympathy with the Past that I could hardly account for, seeing that I was Colonial born and bred. For the first time since my arrival in England the real beauty of the place came home upon me. I felt as if I could have looked for ever on that quiet and peaceful spot.

When we reached the bottom of the hill, and had turned the corner, a broad, well-made stone bridge confronted us. On the other side of this was an old-fashioned country inn, with its signboard dangling from the house front, and opposite it again a dilapidated cottage lolling beside two iron gates. The gates were eight feet or more in height, made of finely wrought iron, and supported by big stone posts, on the top of which two stone animals, griffins, I believe they are called, holding shields in their claws, looked down on passers-by in ferocious grandeur. From behind the gates an avenue wound and disappeared into the wood.

Without consulting me, my old charioteer drove into the inn yard, and, having thrown the reins to an ostler, descended from the vehicle. I followed his example, and then enquired the name of the place inside the gates. My guide, philosopher, and friend looked at me rather queerly for a second or two, and then recollecting that I was a stranger to the place, said:

'That be the Hall I was telling 'ee about. That's where Sir William lives!'

'Then that's where my father was born?'

He nodded his head, and as he did so I noticed that the ostler stopped his work of unharnessing the horse, and looked at me in rather a surprised fashion.

'Well, that being so,' I said, taking my stick from the trap, and preparing to stroll off, 'I'm just going to investigate a bit. You bring yourself to an anchor in yonder, my friend, and don't stir till I come for you again.'

He took himself into the inn without more ado, and I crossed the road towards the gates. They were locked, but the little entrance by the tumble-down cottage stood open, and passing through this I started up the drive. It was a perfect afternoon; the sunshine straggled in through the leafy canopy overhead and danced upon my path. To the right were the thick fastnesses of the preserves; while on my left, across the meadows I could discern the sparkle of water on a weir. I must have proceeded for nearly a mile through the wood before I caught sight of the house. Then, what a strange experience was mine.

I Visit my Relations

Leaving the shelter of the trees, I opened on to as beautiful a park as the mind of man could imagine. A herd of deer were grazing quietly just before me, a woodman was eating his dinner in the shadow of an oak; but it was not upon deer or woodman that I looked, but at the house that stared at me across the undulating sea of grass.

It was a noble building, of grey stone, in shape almost square, with many curious buttresses and angles. The drive ran up to it with a grand sweep, and upon the green that fronted it some big trees reared their stately heads. In my time I'd heard a lot of talk about the stately homes of England, but this was the first time I had ever set eyes on one. And to think that this was my father's birthplace, the house where my ancestors had lived for centuries! I could only stand and stare at it in sheer amazement.

You see, my father had always been a very silent man, and though he used sometimes to tell us yarns about scrapes he'd got into as a boy, and how his father was a very stern man, and had sent him to a public school, because his tutor found him unmanageable, we never thought that he'd been anything very much in the old days—at any rate, not one of such a family as owned this house.

To tell the truth, I felt a bit doubtful as to what I'd better do. Somehow I was rather nervous about going up to the house and introducing myself as a member of the family without any credentials to back my assertion up; and yet, on the other hand, I did not want to go away and have it always rankling in my mind that I'd seen the old place and been afraid to go inside. My mind once made up, however, off I went, crossed the park, and made towards the front door. On nearer approach, I discovered that everything showed the same neglect I had noticed at the lodge. The drive was overgrown with weeds; no carriage seemed to have passed along it for ages. Shutters enclosed many of the windows, and where they did not, not one but several of the panes were broken. Entering the great stone porch, in which it would have been possible to seat a score of people, I pulled the antique doorbell, and waited, while the peal re-echoed down the corridors, for the curtain to go up on the next scene in my domestic drama.

Presently I heard footsteps approaching. A key turned in the lock, and the great door swung open. An old man, whose years could hardly have totalled less than seventy years, stood before me, dressed in a suit of solemn black, almost green with age. He enquired my business in a wheezy whisper. In reply I asked if Sir William Hatteras were at home. Informing me that he would find out, he left me to cool my heels where I stood, and to ruminate on the queerness of my position. In five minutes or so he returned, and signed to me to follow him.

The hall was in keeping with the outside of the building, lofty and imposing. The floor was of oak, almost black with age, the walls were beautifully wainscoted and carved, and here and there tall armoured figures looked down upon me in disdainful silence. But the crowning glory of all was the magnificent staircase that ran up from the centre. It was wide enough and strong enough to have taken a coach and four, the pillars that supported it were exquisitely carved, as were the banisters and rails. Half-way up was a sort of landing, from which again the stairs branched off to right and left.

Above this landing-place, and throwing a stream of coloured light down into the hall, was a magnificent stained-glass window, and on a lozenge in the centre of it the arms that had so much puzzled me on the gateway. A nobler hall no one could wish to possess, but brooding over it was the same air of poverty and neglect I had noticed all about the place. By the time I had taken in these things, my guide had reached a door at the further end. Pushing it open he bade me enter, and I did so, to find a tall, elderly man of stern aspect awaiting my coming.

He, like his servant, was dressed entirely in black, with the exception of a white tie, which gave his figure a semi-clerical appearance. His face was long and somewhat pinched, his chin and upper lip were shaven, and his snow-white, close-cropped whiskers ran in two straight lines from his jaw up to level with his piercing, hawk-like eyes. He would probably have been about seventy-five years of age, but he did not carry it well. In a low, monotonous voice he bade me welcome, and pointed to a chair, himself remaining standing.

'My servant tells me you say your name is Hatteras?' he began.

'That is so,' I replied. 'My father was James Dymoke Hatteras.'

He looked at me very sternly for almost a minute, not for a second betraying the slightest sign of surprise. Then putting his hands together, finger tip to finger tip, as I discovered later was his invariable habit while thinking, he said solemnly:

'James was my younger brother. He misconducted himself gravely in England and was sent abroad. After a brief career of spendthrift extravagance in Australia, we never heard of him again. You may be

I Visit my Relations

his son, but then, on the other hand, of course, you may not. I have no means of judging.'

'I give you my word,' I answered, a little nettled by his speech and the insinuation contained in it; 'but if you want further proof, I've got a Latin book in my portmanteau with my father's name upon the flyleaf, and an inscription in his own writing setting forth that it was given by him to me.'

'A Catullus?'

'Exactly! a Catullus.'

'Then I'll have to trouble you to return it to me at your earliest convenience. The book is my property: I paid eighteenpence for it about eleven o'clock a.m. on the 3rd of July, 1833, in the shop of John Burns, Fleet Street, London. My brother took it from me a week later, and I have not been able to afford myself another copy since.'

'You admit then that the book is evidence of my father's identity?'

'I admit nothing. What do you want with me? What do you come here for? You must see for yourself that I am too poor to be of any service to you, and I have long since lost any public interest I may once have possessed.'

'I want neither one nor the other. I am home from Australia on a trip, and I have a sufficient competence to render me independent of anyone.'

'Ah! That puts a different complexion on the matter. You say you hail from Australia? And what may you have been doing there?'

'Gold-mining—pearling—trading!'

He came a step closer, and as he did so I noticed that his face had assumed a look of indescribable cunning that was evidently intended to be of an ingratiating nature. He spoke in little jerks, pressing his fingers together between each sentence.

'Gold-mining! Ah! And pearling! Well, well! And I suppose you have been fortunate in your ventures?'

'Very!' I replied, having by this time determined on my line of action. 'I daresay my cheque for ten thousand pounds would not be dishonoured by the Bank of England.'

'Ten thousand pounds! Ten thousand pounds! Dear me, dear me!'

He shuffled up and down the dingy room, all the time looking at me out of the corners of his eyes, as if to make sure that I was telling him the truth.

'Come, come, uncle,' I said, resolving to bring him to his bearings without further waste of time. 'This is not a very genial welcome to the son of a long-lost brother!'

'Well, well, you mustn't expect too much, my boy! You see for yourself the position I'm in. The old place is shut up, going to rack and ruin. Poverty is staring me in the face; I am cheated by everybody. Robbed right and left, not knowing which way to turn. But I'll not be put upon. They may call me what they please, but they can't get blood out of a stone. Can they? Answer me that, now!'

This speech showed me everything as plain as a pikestaff. I mean, of course, the reason of the deserted and neglected house, and his extraordinary reception of myself. I rose to my feet.

'Well, uncle—for my uncle you certainly are, whatever you may say to the contrary—I must be going. I'm sorry to find you like this, and from what you tell me I couldn't think of worrying you with my society! I want to see the old church and have a talk with the parson, and then I shall go off never to trouble you again.'

He immediately became almost fulsome in his effort to detain me.

'No, no! You mustn't go like that. It's not hospitable. Besides, you mustn't talk with parson. He's a bad lot is parson—a hard man with a cruel tongue. Says terrible things about me does parson. But I'll be even with him yet. Don't speak to him, laddie, for the honour of the family. Now ye'll stay and take lunch with me?—potluck, of course—I'm too poor to give ye much of a meal; and in the meantime I'll show ye the house and estate.'

This was just what I wanted, though I did not look forward to the prospect of lunch in his company.

With trembling hands he took down an old-fashioned hat from a peg and turned towards the door. When we had passed through it he carefully locked it and dropped the key into his breeches' pocket. Then he led the way upstairs by the beautiful oak staircase I had so much admired on entering the house.

When we reached the first landing, which was of noble proportions and must have contained upon its walls nearly a hundred family portraits all coated with the dust of years, he approached a door and threw it open. A feeble light straggled in through the closed shutters, and revealed an almost empty room. In the centre stood a large canopied bed, of antique design. The walls were wainscotted, and the massive chimney-piece was carved with heraldic designs. I enquired what room this might be.

'This is where all our family were born,' he answered. ' 'Twas here your father first saw the light of day.'

I looked at it with a new interest. It seemed hard to believe that this was the birthplace of my own father, the man whom I remembered so well in a place and life so widely different. My companion noticed the look upon my face, and, I suppose, felt constrained to say something.

'Ah! James!' he said sorrowfully, 'ye were always a giddy, roving lad. I remember ye well.' (He passed his hand across his eyes, to brush away a tear, I thought, but his next speech disabused me of any such notion.) 'I remember that but a day or two before ye went ye blooded my nose in the orchard, and the very morning ye decamped ye borrowed half a crown of me, and never paid it back.'

A sudden something prompted me to put my hand in my pocket. I took out half a crown, and handed it to him without a word. He took it, looked at it longingly, put it in his pocket, took it out again, ruminated a moment, and then reluctantly handed it back to me.

'Nay, nay! my laddie, keep your money, keep your money. Ye can send me the Catullus.' Then to himself, unconscious that he was speaking his thoughts aloud: 'It was a good edition, and I have no doubt would bring five shillings any day.'

From one room we passed into another, and yet another. They were all alike—shut up, dust-ridden, and forsaken. And yet with it all what a noble place it was—one which any man might be proud to call his own. And to think that it was all going to rack and ruin because of the miserly nature of its owner. In the course of our ramble I discovered that he kept but two servants, the old man who had admitted me to his presence, and his wife, who, as that peculiar phrase has it, cooked and did for him. I discovered later that he had not paid either of them wages for some years past, and that they only stayed on with him because they were too poor and proud to seek shelter elsewhere.

When we had inspected the house we left it by a side door, and crossed a courtyard to the stables. There the desolation was, perhaps, even more marked than in the house. The great clock on the tower above the main building had stopped at a quarter to ten on some long-forgotten day, and a spider now ran his web from hand to hand.

At our feet, between the stones, grass grew luxuriantly, thick moss covered the coping of the well, the doors were almost off their hinges, and rats scuttled through the empty loose boxes at our approach. So large was the place, that thirty horses might have found a lodging comfortably, and as far as I could gather, there was room for half as many vehicles in the coach-houses that stood on either side. The intense quiet was only broken by the cawing of the rooks in the giant elms overhead, the squeaking of the rats, and the low grumbling of my uncle's voice as he pointed out the ruin that was creeping over everything.

Before we had finished our inspection it was lunch time, and we returned to the house. The meal was served in the same room in which I had made my relative's acquaintance an hour before. It consisted, I discovered, of two meagre mutton chops and some home-made bread and cheese, plain and substantial fare enough in its way, but hardly the sort one would expect from the owner of such a house. For a beverage, water was placed before us, but I could see that my host was deliberating as to whether he should stretch his generosity a point or two further.

Presently he rose, and with a muttered apology left the room, to return five minutes later carrying a small bottle carefully in his hand. This, with much deliberation and no small amount of sighing, he opened. It proved to be claret, and he poured out a glassful for me. As I was not prepared for so much liberality, I thought something must be behind it, and in this I was not mistaken.

'Nephew,' said he after a while, 'was it ten thousand pounds you mentioned as the amount of your fortune?'

I nodded. He looked at me slyly and cleared his throat to gain time for reflection. Then seeing that I had emptied my glass, he refilled it with another scarce concealed sigh, and sat back in his chair.

'And I understand you to say you are quite alone in the world, my boy?'

'Quite! Until I met you this morning I was unaware that I had a single relative on earth. Have I any more connections?'

'Not a soul—only Gwendoline.'

'Gwendoline!' I cried, 'and who may Gwendoline be?'

'My daughter—your cousin. My only child! Would you like to see her?'

'I had no idea you had a daughter. Of course I should like to see her!'

He left the table and rang the bell. The ancient man-servant answered the summons.

'Tell your wife to bring Miss Gwendoline to us.'

'Miss Gwendoline here, sir? You do not mean it sure-lie, sir?'

'Numbskull! numbskull! numbskull!' cried the old fellow in an ecstasy of fury that seemed to spring up as suddenly as a squall does between the islands, 'bring her without another word or I'll be the death of you.'

Without further remonstrance the old man left the room, and I demanded an explanation.

'Good servant, but an impudent rascal, sir!' he said. 'Of course you must see my daughter, my beautiful daughter, Gwendoline. He's afraid you'll frighten her, I suppose! Ha! ha! Frighten my bashful, pretty one. Ha! ha!'

Anything so supremely devilish as the dried-up mirth of this old fellow it would be difficult to imagine. His very laugh seemed as if it had to crack in his throat before it could pass his lips. What would his daughter be like, living in such a house, with such companions? While I was wondering, I heard footsteps in the corridor, and then an old woman entered and curtsied respectfully. My host rose and went over to the fireplace, where he stood with his hands behind his back and the same devilish grin upon his face.

'Well, where is my daughter?'

'Sir, do you really mean it?'

'Of course, I mean it. Where is she?'

In answer the old lady went to the door and called to someone in the hall.

'Come in, dearie. It's all right. Come in, do'ee now, that's a little dear.'

But the girl made no sign of entering, and at last the old woman had to go out and draw her in. And then—but I hardly know how to write it. How shall I give you a proper description of the—*thing* that entered.

She—if *she* it could be called—was about three feet high, dressed in a shapeless print costume. Her hair stood and hung in a tangled mass upon her head, her eyes were too large for her face, and to complete the horrible effect, a great patch of beard grew on one cheek, and descended almost to a level with her chin. Her features were all awry, and now and again she uttered little moans that were more like those of a wild beast than of a human being. In spite of the old woman's endeavours to make her do so, she would not venture from her side, but stood slobbering and moaning in the half dark of the doorway.

It was a ghastly sight, one that nearly turned me sick with loathing. But the worst part of it all was the inhuman merriment of her father.

'There, there!' he cried; 'had ever man such a lovely daughter? Isn't she a beauty? Isn't she fit to be a prince's bride? Isn't she fit to be the heiress of all this place? Won't the young dukes be asking her hand in marriage? Oh, you beauty! You—but there, take her away—take her away, I say, before I do her mischief.'

The words had no sooner left his mouth than the old woman seized her charge and bundled her out of the room, moaning as before. I can tell you there was at least one person in that apartment who was heartily glad to be rid of her.

When the door had closed upon them my host came back to his seat, and with another sigh refilled my glass. I wondered what was coming next. It was not long, however, before I found out.

'Now you know everything,' he said. 'You have seen my home, you have seen my poverty, and you have seen my daughter. What do you think of it all?'

'I don't know what to think.'

'Well, then, I'll tell you. That child wants doctors; that child wants proper attendance. She can get neither here. I am too poor to help her in any way. You're rich by your own telling. I have today taken you into the bosom of my family, recognized you without doubting your assertions. Will you help me? Will you give me one thousand pounds towards settling that child in life? With that amount it could be managed.'

'Will I what?' I cried in utter amazement—dumbfounded by his impudence.

'Will you settle one thousand pounds upon her, to keep her out of her grave?'

'Not one penny!' I cried; 'and, what's more, you miserable, miserly old wretch, I'll give you a bit of my mind.'

And thereupon I did! Such a talking to as I suppose the old fellow had never had in his life before, and one he'd not be likely to forget

I Visit my Relations

in a hurry. He sat all the time, white with fury, his eyes blazing, and his fingers quivering with impotent rage. When I had done he ordered me out of his house. I took him at his word, seized my hat, and strode across the hall through the front door, and out into the open air.

But I was not to leave the home of my ancestors without a parting shot. As I closed the front door behind me I heard a window go up, and on looking round there was the old fellow shaking his fist at me from the second floor.

'Leave my house—leave my park!' he cried in a shrill falsetto, 'or I'll send for the constable to turn you off. Bah! You came to steal. You're no nephew of mine; I disown you! You're a common cheat—a swindler—an impostor! Go!'

I took him at his word, and went. Leaving the park, I walked straight across to the rectory, and enquired if I might see the clergyman. To him I told my tale, and, among other things, asked if anything could be done for the child—my cousin. He only shook his head.

'I fear it is hopeless, Mr Hatteras,' the clergyman said. 'The old gentleman is a terrible character, and as he owns half the village, and every acre of the land hereabouts, we all live in fear and trembling of him. We have no shadow of a claim upon the child, and unless we can prove that he actually ill-treats it, I'm sorry to say I think there is nothing to be done.'

So ended my first meeting with my father's family.

From the rectory I returned to my inn. What should I do now? London was worse than a desert to me now that my sweetheart was gone from it, and every other place seemed as bad. Then an advertisement on the wall of the bar parlour caught my eye:

FOR SALE OR HIRE,
THE YACHT, 'ENCHANTRESS.'
Ten Tons.
Apply, SCREW & MATCHEM, Bournemouth.

It was just the very thing. I was pining for a breath of sea air again. It was perfect weather for a cruise. I would go to Bournemouth, inspect the yacht at once, and, if she suited me, take her for a month or so. My mind once made up, I hunted up my Jehu, and set off for the train, never dreaming that by so doing I was taking the second step in that important chain of events that was to affect all the future of my life.

CHAPTER IV
I Save an Important Life

To a man whose life has been spent in the uttermost parts of the earth, amid barbaric surroundings, and in furtherance of work of a kind that the civilized world usually denominates dangerous, the seaside life of England must afford scope for wonderment and no small amount of thoughtful consideration. And certainly if there is one place more than another where, winter and summer alike, amid every sort of luxury, the modern Englishman may be seen relaxing his cares and increasing his energies, the name of that place is Bournemouth. Built up amid pine-woods, its beauties added to in every fashion known to the fertile brain of man, Bournemouth stands, to my mind, pre-eminent in the list of British watering-places.

Leaving Lyndhurst Road, I travelled to this excellent place by a fast train, and immediately on arrival made my way to the office of Messrs Screw & Matchem with a view to instituting enquiries regarding the yacht they had advertised for hire. It was with the senior partner I transacted my business, and a shrewd but pleasant gentleman I found him.

Upon my making known my business to him, he brought me a photograph of the craft in question, and certainly a nice handy boat she looked. She had been built, he went on to inform me, for a young nobleman, who had made two very considerable excursions in her before he had been compelled to fly the country, and was only three years old. I learned also that she was lying in Poole harbour, but he was good enough to say that if I wished to see her she should be brought round to Bournemouth the following morning, when I could inspect her at my leisure. As this arrangement was one that exactly suited me, I closed with it there and then, and thanking Mr Matchem for his courtesy, betook myself to my hotel. Having dined, I spent the evening upon the pier—the first of its kind I had ever seen—listened to the band, and diverted myself with thoughts of her to whom I had plighted my troth, and whose unexpected departure from England had been such a sudden and bitter disappointment to me.

Next morning, faithful to promise, the *Enchantress* sailed into the bay and came to an anchor within a biscuit throw of the pier. Chartering a dinghy, I pulled myself off to her, and stepped aboard. An old man and a boy were engaged washing down, and to them I introduced myself and my business. Then for half an hour I devoted myself to overhauling her thoroughly. She was a nice enough little craft, well set up, and from her run looked as if she might possess a fair turn of speed; the gear was in excellent order, and this was accounted for when the old man told me she had been repaired and thoroughly overhauled that selfsame year.

Having satisfied myself on a few other minor points, I pulled ashore and again went up through the gardens to the agents' office. Mr Matchem was delighted to hear that I liked the yacht well enough to think of hiring her at their own price (a rather excessive one, I must admit), and, I don't doubt, would have supplied me with a villa in Bournemouth, and a yachting box in the Isle of Wight, also on their own terms, had I felt inclined to furnish them with the necessary order. But fortunately I was able to withstand their temptations, and having given them my cheque for the requisite amount, went off to make arrangements, and to engage a crew.

Before nightfall I had secured the services of a handy lad in place of the old man who had brought the boat round from Poole, and was in a position to put to sea. Accordingly next morning I weighed anchor for a trip round the Isle of Wight. Before we had brought the Needles abeam I had convinced myself that the boat was an excellent sailer, and when the first day's cruise was over I had found no reason to repent having hired her.

And I would ask you here, is there any other amusement to compare with yachting? Can anything else hope to vie with it? Suppose a man to be a lover of human craftmanship—then what could be more to his taste than a well-built yacht? Is a man a lover of speed? Then what more could he wish for than the rush over the curling seas, with the graceful fabric quivering under him like an eager horse, the snowy line of foam driving away from either bow, and the fresh breeze singing merrily through the shrouds, bellying out the stretch of canvas overhead till it seems as if the spars must certainly give way beneath the strain they are called upon to endure!

Is a man a lover of the beautiful in nature? Then from what better place can he observe earth, sky, and sea than from a yacht's deck? Thence he views the stretch of country ashore, the dancing waves around him, the blue sky flaked with fleecy clouds above his head, while the warm sunshine penetrates him through and through till it finds his very heart and stays there, making a better, and certainly a healthier, man of him.

Does the world ever look so fair as at daybreak, when Dame Nature is still half asleep, and the water lies like a sheet of shimmering glass, and the great sun comes up like a ball of gold, with a solemnity that makes one feel almost afraid? Or at night when, anchored in some tiny harbour, the lights are twinkling ashore, and the sound of music comes wafted across the water, with a faintness that only adds to its beauty, to harmonize with the tinkling of the waves alongside. Review these things in your mind and then tell me what recreation can compare with yachting?

Not having anything to hurry me, and only a small boy and my own thoughts to keep my company, I took my time; remained two days in the Solent, sailed round the island, put in a day at Ventnor, and so back to Bournemouth. Then, after a day ashore, I picked up a nice breeze and ran down to Torquay to spend another week sailing slowly back along the coast, touching at various ports, and returning eventually to the place I had first hailed from.

In relating these trifling incidents it is not my wish to bore my readers, but to work up gradually to that strange meeting to which they were the prelude. Now that I can look back in cold blood upon the circumstances that brought it about, and reflect how narrowly I escaped missing the one event which was destined to change my whole life, I can hardly realize that I attached such small importance to it at the time. Somehow I have always been a firm believer in Fate, and indeed it would be strange, all things considered, if I were not. For when a man has passed through so many extraordinary adventures as I have, and not only come out of them unharmed, but happier and a great deal more fortunate than he has really any right to be, he may claim the privilege, I think, of saying he knows something about his subject.

I Save an Important Life

And, mind you, I date it all back to that visit to the old home, and to my uncle's strange reception of me, for had I not gone down into the country I should never have quarrelled with him, and if I had not quarrelled with him I should not have gone back to the inn in such a dudgeon, and in that case I should probably have left the place without a visit to the bar, never have seen the advertisement, visited Bournemouth, hired the yacht or—but there I must stop. You must work out the rest for yourself when you have heard my story.

The morning after my third return to Bournemouth I was up by daybreak, had had my breakfast, and was ready to set off on a cruise across the bay, before the sun was a hand's breath above the horizon. It was as perfect a morning as any man could wish to see. A faint breeze just blurred the surface of the water, tiny waves danced in the sunshine, and my barkie nodded to them as if she were anxious to be off. The town ashore lay very quiet and peaceful, and so still was the air that the cries of a few white gulls could be heard quite distinctly, though they were half a mile or more away. Having hove anchor, we tacked slowly across the bay, passed the pier-head, and steered for Old Harry Rock and Swanage Bay. My crew was for'ard, and I had possession of the tiller.

As we went about between Canford Cliffs and Alum Chine, something moving in the water ahead of me attracted my attention. We were too far off to make out exactly what it might be, and it was not until five minutes later, when we were close abreast of it, that I discovered it to be a bather. The foolish fellow had ventured further out than was prudent, had struck a strong current, and was now being washed swiftly out to sea. But for the splashing he made to show his whereabouts, I should in all probability not have seen him, and in that case his fate would have been sealed. As it was, when we came up with him he was quite exhausted.

Heaving my craft to, I leapt into the dinghy, and pulled towards him, but before I could reach the spot he had sunk. At first I thought he was gone for good and all, but in a few seconds he rose again. Then, grabbing him by the hair, I passed an arm under each of his, and dragged him unconscious into the boat. In less than three minutes we were alongside the yacht again, and with my crew's assistance I got him aboard. Fortunately a day or two before I had had the forethought to purchase some brandy for use in case of need, and my Thursday Island experiences having taught me exactly what was best to be done under such circumstances, it was not long before I had brought him back to consciousness.

In appearance he was a handsome young fellow, well set up, and possibly nineteen or twenty years of age. When I had given him a stiff nobbler of brandy to stop the chattering of his teeth, I asked him how he came to be so far from shore.

'I am considered a very good swimmer,' he replied, 'and often come out as far as this, but today I think I must have got into a strong outward current, and certainly but for your providential assistance I should never have reached home alive.'

'You have had a very narrow escape,' I answered, 'but thank goodness you're none the worse for it. Now, what's the best thing to be done? Turn back, I suppose, and set you ashore.'

'But what a lot of trouble I'm putting you to.'

'Nonsense! I've nothing to do, and I count myself very fortunate in having been able to render you this small assistance. The breeze is freshening, and it won't take us any time to get back. Where do you live?'

'To the left there! That house standing back upon the cliff. Really I don't know how to express my gratitude.'

'Just keep that till I ask you for it; and now, as we've got a twenty minutes' sail before us, the best thing for you to do would be to slip into a spare suit of my things. They'll keep you warm, and you can return them to my hotel when you get ashore.'

I sang out to the boy to come aft and take the tiller, while I escorted my guest below into the little box of a cabin, and gave him a rig out. Considering I am six feet two, and he was only five feet eight, the things were a trifle large for him; but when he was dressed I couldn't help thinking what a handsome, well-built, aristocratic-looking young fellow he was. The work of fitting him out accomplished, we returned to the deck. The breeze was freshening, and the little hooker was ploughing her way through it, nose down, as if she knew that under the circumstances her best was expected of her.

'Are you a stranger in Bournemouth?' my companion asked as I took the tiller again.

'Almost,' I answered. 'I've only been in England three weeks. I'm home from Australia.'

'Australia! Really! Oh, I should so much like to go out there.'

His voice was very soft and low, more like a girl's than a boy's, and I noticed that he had none of the mannerisms of a man—at least, not of one who has seen much of the world.

'Yes, Australia's as good a place as any other for the man who goes out there to work,' I said, 'But somehow you don't look to me like a chap that is used to what is called roughing it. Pardon my bluntness.'

'Well, you see, I've never had much chance. My father is considered by many a very peculiar man. He has strange ideas about me, and so you see I've never been allowed to mix with other people. But I'm stronger than you'd think, and I shall be twenty in October next.'

I wasn't very far out in his age then.

'And now, if you don't mind telling me, what is your name?'

'I suppose there can be no harm in letting you know it. I was told if ever I met anyone and they asked me, not to tell them. But since you saved my life it would be ungrateful not to let you know. I am the Marquis of Beckenham.'

'Is that so? Then your father is the Duke of Glenbarth?'

'Yes. Do you know him?'

'Never set eyes on him in my life, but I heard him spoken of the other day.'

I did not add that it was Mr Matchem who, during my conversation with him, had referred to his Grace, nor did I think it well to say that he had designated him the 'Mad Duke'. And so the boy I had saved from drowning was the young Marquis of Beckenham. Well, I was moving in good society with a vengeance. This boy was the first nobleman I had ever clapped eyes on, though I knew the Count de Panuroff well enough in Thursday Island. But then foreign Counts, and shady ones at that, ought not to reckon, perhaps.

'But you don't mean to tell me,' I said at length, 'that you've got no friends? Don't you ever see anyone at all?'

'No, I am not allowed to. My father thinks it better not. And as he does not wish it, of course I have nothing left but to obey. I must own, however, I should like to see the world—to go a long voyage to Australia, for instance.'

'But how do you put in your time? You must have a very dull life of it.'

'Oh, no! You see, I have never known anything else, and then I have always the future to look forward to. As it is now, I bathe every morning, I have my yacht, I ride about the park, I have my studies, and I have a tutor who tells me wonderful stories of the world.'

'Oh, your tutor has been about, has he?'

'Dear, yes! He was a missionary in the South Sea Islands, and has seen some very stirring adventures.'

'A missionary in the South Seas, eh? Perhaps I know him.'

'Were you ever in those seas?'

'Why, I've spent almost all my life there.'

'Were you a missionary?'

'You bet not. The missionaries and my friends don't cotton to one another.'

'But they are such good men!'

'That may be. Still, as I say, we don't somehow cotton. D'you know I'd like to set my eyes upon your tutor.'

'Well, you will. I think I see him on the beach now. I expect he has been wondering what has become of me. I've never been out so long before.'

'Well, you're close home now, and as safe as eggs in a basket.'

Another minute brought us into as shallow water as I cared to go. Accordingly, heaving to, I brought the dinghy alongside, and we got into her. Then casting off, I pulled my lord ashore. A small, clean-shaven, parsonish-looking man, with the regulation white choker, stood by the water waiting for us. As I beached the boat he came forward and said:

'My lord, we have been very anxious about you. We feared you had met with an accident.'

'I have been very nearly drowned, Mr Baxter. Had it not been for this gentleman's prompt assistance I should never have reached home again.'

'You should really be more careful, my lord. I have warned you before. You father has been nearly beside himself with anxiety about you!'

'Eh?' said I to myself. 'Somehow this does not sound quite right. Anyhow, Mr Baxter, I've seen your figure-head somewhere before—but you were not a missionary then, I'll take my affidavit.'

Turning to me, my young lord held out his hand.

'You have never told me your name,' he said almost reproachfully.

'Dick Hatteras,' I answered, 'and very much at your service.'

'Mr Hatteras, I shall never forget what you have done for me. That I am most grateful to you I hope you will believe. I know that I owe you my life.'

Here the tutor's voice chipped in again, as I thought, rather impatiently.

'Come, come, my lord. This delay will not do. Your father will be growing still more nervous about you. We must be getting home!'

Then they went off up the cliff path together, and I returned to my boat.

'Mr Baxter,' I said to myself again as I pulled off to the yacht, 'I want to know where I've seen your face before. I've taken a sudden dislike to you. I don't trust you; and if your employer's the man they say he is, well, he won't either.'

Then, having brought the dinghy alongside, I made the painter fast, clambered aboard, and we stood out of the bay once more.

CHAPTER V
Mystery

The following morning I was sitting in my room at the hotel idly scanning the *Standard*, and wondering in what way I should employ myself until the time arrived for me to board the yacht, when I heard a carriage roll up to the door.

On looking out I discovered a gorgeous landau, drawn by a pair of fine thoroughbreds, and resplendent with much gilded and crested harness, standing before the steps. A footman had already opened the door, and I was at the window just in time to see a tall, soldierly man alight from it. To my astonishment, two minutes later a waiter entered my room and announced 'His Grace the Duke of Glenbarth'. It was the owner of the carriage and the father of my young friend, if by such a title I might designate the Marquis of Beckenham.

'Mr Hatteras, I presume?' said he, advancing towards me and using that dignified tone that only an English gentleman can assume with anything approaching success.

'Yes, that is my name. I am honoured by your visit. Won't you sit down?'

'Thank you.'

He paused for a moment, and then continued:

'Mr Hatteras, I have to offer you an apology. I should have called upon you yesterday to express the gratitude I feel to you for having saved the life of my son, but I was unavoidably prevented.'

'I beg you will not mention it,' I said. 'His lordship thanked me sufficiently himself. And after all, when you look at it, it was not very much to do. I would, however, venture one little suggestion. Is it wise to let him swim so far unaccompanied by a boat? The same thing might happen to him on another occasion, and no one be near enough to render him any assistance.'

'He will not attempt so much again. He has learned a lesson from this experience. And now, Mr Hatteras, I trust you will forgive what I am about to say. My son has told me that you have just arrived in England from Australia. Is there any way I can be of service to you? If there is, and you will acquaint me with it, you will be conferring a great favour upon me.'

'I thank your Grace,' I replied—I hope with some little touch of dignity—'it is very kind of you, but I could not think of such a thing. But, stay, there is one service perhaps you *could* do me.'

'I am delighted to hear it, sir. And pray what may it be?'

'Your son's tutor, Mr Baxter! His face is strangely familiar to me. I have seen him somewhere before, but I cannot recall where. Could you tell me anything of his history?'

'Very little, I fear, save that he seems a worthy and painstaking man, an excellent scholar, and very capable in his management of young men. I received excellent references with him, but of his past history I know very little. I believe, however, that he was a missionary in the South Seas for some time, and that he was afterwards for many years in India. I'm sorry I cannot tell you more about him since you are interested in him.'

'I've met him somewhere, I'm certain. His face haunts me. But to return to your son—I hope he is none the worse for his adventure?'

'Not at all, thank you. Owing to the system I have adopted in his education, the lad is seldom ailing.'

'Pardon my introducing the subject. But do you think it is quite wise to keep a youth so ignorant of the world? I am perhaps rather presumptuous, but I cannot help feeling that such a fine young fellow would be all the better for a few companions.'

'You hit me on rather a tender spot, Mr Hatteras. But, as you have been frank with me, I will be frank with you. I am one of those strange beings who govern their lives by theories. I was brought up by my father, I must tell you, in a fashion totally different from that I am employing with my son. I feel now that I was allowed a dangerous amount of license. And what was the result? I mixed with everyone, was pampered and flattered far beyond what was good for me, derived a false notion of my own importance, and when I came to man's estate was, to all intents and purposes, quite unprepared and unfitted to undertake the duties and responsibilities of my position.

'Fortunately I had the wit to see where the fault lay, and there and then I resolved that if ever I were blessed with a son, I would conduct his education on far different lines. My boy has not met a dozen strangers in his life. His education has been my tenderest care. His position, his duties towards his fellow-men, the responsibilities of his rank, have always been kept rigorously before him. He has been brought up to understand that to be a Duke is not to be a titled nonentity or a pampered roué, but to be one whom Providence has blessed with an opportunity of benefiting and watching over the welfare of those less fortunate than himself in the world's good gifts.

'He has no exaggerated idea of his own importance; a humbler lad, I feel justified in saying, you would nowhere find. He has been educated thoroughly, and he has all the best traditions of his race kept continually before his eyes. But you must not imagine, Mr Hatteras, that because he has not mixed with the world he is ignorant of its temptations. He may not have come into personal contact with them, but he has been warned against their insidious influences, and I shall trust to his personal pride and good instincts to help him to withstand them when he has to encounter them himself. Now, what do you think of my plan for making a nobleman?'

'A very good one, with such a youth as your son, I should think, your Grace; but I would like to make one more suggestion, if you would allow me?'

'And that is?'

'That you should let him travel before he settles down. Choose some fit person to accompany him. Let him have introductions to good people abroad, and let him use them; then he will derive different impressions from different countries, view men and women from different standpoints, and enter gradually into the great world and station which he is some day to adorn.'

'I had thought of that myself, and his tutor has lately spoken to me a good deal upon the subject. I must own it is an idea that commends itself strongly to me. I will think it over. And now, sir, I must wish you good-day. You will not let me thank you, as I should have wished, for the service you have rendered my house, but, believe me, I am none the less grateful. By the way, your name is not a common one. May I ask if you have any relatives in this county?'

'Only one at present, I fancy—my father's brother, Sir William Hatteras, of Murdlestone, in the New Forest.'

'Ah! I never met him. I knew his brother James very well in my younger days. But he got into sad trouble, poor fellow, and was obliged to fly the country.'

'You are speaking of my father. And you knew him?'

'Knew him? indeed, I did. And a better fellow never stepped; but, like most of us in those days, too wild—much too wild! And so you are James's son? Well, well! This is indeed a strange coincidence. But dear me, I am forgetting; I must beg your pardon for speaking so candidly of your father.'

'No offence, I'm sure.'

'And pray tell me where my old friend is now?'

'Dead, your Grace! He was drowned at sea.'

The worthy old gentleman seemed really distressed at this news. He shook his head, and I heard him murmur:

'Poor Jim! Poor Jim!'

Then, turning to me again, he took my hand.

'This makes our bond a doubly strong one. You must let me see more of you! How long do you propose remaining in England?'

'Not very much longer, I fear. I am already beginning to hunger for the South again.'

'Well, you must not go before you have paid us a visit. Remember we shall always be pleased to see you. You know our house, I think, on the cliff. Good-day, sir, good-day.'

So saying, the old gentleman accompanied me downstairs to his carriage, and, shaking me warmly by the hand, departed.

Again I had cause to ponder on the strangeness of the fate that had led me to Hampshire—first to the village where my father was born, and then to Bournemouth, where by saving this young man's life I had made a firm friend of a man who again had known my father. By such small coincidences are the currents of our lives diverted.

That same afternoon, while tacking slowly down the bay, I met the Marquis. He was pulling himself in a small skiff, and when he saw me he made haste to come alongside and hitch on. At first I wondered whether it would not be against his father's wishes that he should enter into conversation with such a worldly person as myself. But he evidently saw what was passing in my mind, and banished all doubts by saying:

'I have been on the look-out for you, Mr Hatteras. My father has given me permission to cultivate your acquaintance, if you will allow me.'

'I shall be very pleased,' I answered. 'Won't you come aboard and have a chat? I'm not going out of the bay this afternoon.'

He clambered over the side and seated himself in the well, clear of the boom, as nice-looking and pleasant a young fellow as any man could wish to set eyes on. 'Well,' I thought to myself, 'if all Peers were like this boy there'd be less talk of abolishing the House of Lords.'

'You can't imagine how I've been thinking over all you told me the other day,' he began when we were fairly on our way. 'I want you to tell me more about Australia and the life you lead out there, if you will.'

'I'll tell you all I can with pleasure,' I answered. 'But you ought to go and see the places and things for yourself. That's better than any telling. I wish I could take you up and carry you off with me now; away down to where you can make out the green islands peeping out of the water to port and starboard, like bits of the Garden of Eden gone astray and floated out to sea. I'd like you to smell the breezes that come off from them towards evening, to hear the "trades" whistling overhead, and the thunder of the surf upon the reef. Or at another time to get inside that selfsame reef and look down through the still, transparent water, at the rainbow-coloured fish dashing among the coral boulders, in and out of the most beautiful fairy grottoes the brain of man can conceive.'

'Oh, it must be lovely! And to think that I may live my life and never see these wonders. Please go on; what else can you tell me?'

'What more do you want to hear? There is the pick of every sort of life for you out there. Would you know what real excitement is? Then

I shall take you to a new gold rush. To begin with you must imagine yourself setting off for the field, with your trusty mate marching step by step beside you, pick and shovel on your shoulders, and both resolved to make your fortunes in the twinkling of an eye. When you get there, there's the digger crowd, composed of every nationality. There's the warden and his staff, the police officers, the shanty keepers, the blacks, and dogs.

'There's the tented valley stretching away to right and left of you, with the constant roar of sluice boxes and cradles, the creak of windlasses, and the perpetual noise of human voices. There's the excitement of pegging out your claim and sinking your first shaft, wondering all the time whether it will turn up trumps or nothing. There's the honest, manly labour from dawn to dusk. And then, when daylight fails, and the lamps begin to sparkle over the field, songs drift up the hillside from the drinking shanties in the valley, and you and your mate weigh up your day's returns, and, having done so, turn into your blankets to dream of the monster nugget you intend to find upon the morrow. Isn't that real life for you?'

He did not answer, but there was a sparkle in his eyes which told me I was understood.

'Then if you want other sorts of enterprise, there is Thursday Island, where I hail from, with its extraordinary people. Let us suppose ourselves wandering down the Front at nightfall, past the Kanaka billiard saloons and the Chinese stores, into, say, the Hotel of All Nations. Who is that handsome, dark, mysterious fellow, smoking a cigarette and idly flirting with the pretty bar girl? *You* don't know him, but I do! There's indeed a history for you. You didn't notice, perhaps, that rakish schooner that came to anchor in the bay early in the forenoon. What lines she had! Well, that was his craft. Tomorrow she'll be gone, it is whispered, to try for pearl in prohibited Dutch waters. Can't you imagine her slinking round the islands, watching for the patrolling gunboat, and ready, directly she has passed, to slip into the bay, skim it of its shell, and put to sea again. Sometimes they're chased—and then?'

'What then?'

'Well, a clean pair of heels or trouble with the authorities, and possibly a year in a Dutch prison before you're brought to trial! Or would you do a pearling trip in less exciting but more honest fashion?

Would you ship aboard a lugger with five good companions, and go a-cruising down the New Guinea coast, working hard all day long, and lying out on deck at night, smoking and listening to the lip-lap of the water against the counter, or spinning yarns of all the world?'

'What else?'

'Why, what more do you want? Do you hanker after a cruise aboard a stinking *bêche-de-mer* boat inside the Barrier Reef, or a run with the sandalwood cutters or tortoiseshell gatherers to New Guinea; or do you want to go ashore again and try an overlanding trip half across the continent, riding behind your cattle all day long, and standing your watch at night under dripping boughs, your teeth chattering in your head, waiting for the bulls to break, while every moment you expect to hear the Bunyip calling in that lonely water-hole beyond the fringe of Mulga scrub?'

'You make me almost mad with longing.'

'And yet, somehow, it doesn't seem so fine when you're at it. It's when you come to look back upon it all from a distance of twelve thousand miles that you feel its real charm. Then it calls to you to return in every rustle of the leaves ashore, in the blue of the sky above, in the ripple of the waves upon the beach. And it eats into your heart, so that you begin to think you will never be happy till you're back in the old tumultuous devil-may-care existence again.'

'What a life you've led! And how much more to be envied it seems than the dull monotony of our existence here in sleepy old England.'

'Don't you believe it. If you wanted to change I could tell you of dozens of men, living exactly the sort of life I've described, who would only too willingly oblige you. No, no! Believe me, you've got chances of doing things we would never dream of. Do them, then, and let the other go. But all the same, I think you ought to see more of the world I've told you of before you settle down. In fact, I hinted as much to your father only yesterday.'

'He said that you had spoken of it to him. Oh, how I wish he would let me go!'

'Somehow, d'you know, I think he will.'

I put the cutter over on another tack, and we went crashing back through the blue water towards the pier. The strains of the band came faintly off to us. I had enjoyed my sail, for I had taken a great fancy to this bright young fellow sitting by my side. I felt I should like to have finished the education his father had so gallantly begun. There was something irresistibly attractive about him, so modest, so unassuming, and yet so straightforward and gentlemanly.

Dropping him opposite the bathing machines, I went on to my own anchorage on the other side of the pier. Then I pulled myself ashore and went up to the town. I had forgotten to write an important letter that morning, and as it was essential that the business should be attended to at once, to repair my carelessness, I crossed the public gardens and went through the gardens to the post office to send a telegram.

I must tell you here that since my meeting with Mr Baxter, the young Marquis's tutor, I had been thinking a great deal about him, and the more I thought the more certain I became that we had met before. To tell the truth, a great distrust of the man was upon me. It was one of those peculiar antipathies that no one can explain. I did not like his face, and I felt sure that he did not boast any too much love for me.

As my thoughts were still occupied with him, my astonishment may be imagined, on arriving at the building, at meeting him face to face upon the steps. He seemed much put out at seeing me, and hummed and hawed over his 'Good-afternoon' for all the world as if I had caught him in the middle of some guilty action.

Returning his salutation, I entered the building and looked about me for a desk at which to write my wire. There was only one vacant, and I noticed that the pencil suspended on the string was still swinging to and fro as it had been dropped. Now Baxter had only just left the building, so there could be no possible doubt that it was he who had last used the stand. I pulled the form towards me and prepared to write. But as I did so I noticed that the previous writer had pressed so hard upon his pencil that he had left the exact impression of his message plainly visible upon the pad. It ran as follows:

LETTER RECEIVED. YOU OMITTED REVEREND. THE TRAIN IS LAID, BUT A NEW ELEMENT OF DANGER HAS ARISEN.

It was addressed to 'Nikola, Green Sailor Hotel, East India Dock Road, London,' and was signed 'Nineveh'.

The message was so curious that I looked at it again, and the longer I looked the more certain I became that Baxter was the sender. Partly because its wording interested me, and partly for another reason which will become apparent later on, I inked the message over, tore it from the pad, and placed it carefully in my pocket-book. One thing at least was certain, and that was, if Baxter *were* the sender, there was something underhand going on. If he were not, well, then there could be no possible harm in my keeping the form as a little souvenir of a rather curious experience.

I wrote my own message, and having paid for it left the office. But I was not destined to have the society of my own thoughts for long. Hardly had I reached the Invalids' Walk before I felt my arm touched. To my supreme astonishment I found myself again confronted by Mr Baxter. He was now perfectly calm and greeted me with extraordinary civility.

'Mr Hatteras, I believe,' he said. 'I think I had the pleasure of meeting you on the sands a few days ago. What a beautiful day it is, isn't it? Are you proceeding this way? Yes? Then perhaps I may be permitted the honour of walking a short distance with you.'

'With pleasure,' I replied. 'I am going up the cliff to my hotel, and I shall be glad of your company. I think we met in the telegraph office just now.'

'In the post office, I think. I had occasion to go in there to register a letter.'

His speech struck me as remarkable. My observation was so trivial that it hardly needed an answer, and yet not only did he vouchsafe me one, but he corrected my statement and volunteered a further one on his own account. What reason could he have for wanting to make me understand that he had gone in there to post a letter? What would it have mattered to me if he *had* been there, as I suggested, to send a telegram?

'Mr Baxter,' I thought to myself, 'I've got a sort of conviction that you're not the man you pretend to be, and what's more I'd like to bet a shilling to a halfpenny that, if the truth were only known, you're our mysterious friend Nineveh.'

We walked for some distance in silence. Presently my companion began to talk again—this time, however, in a new strain and perhaps with a little more caution.

'You have been a great traveller, I understand, Mr Hatteras.'

'A fairly great one, Mr Baxter. You also, I am told, have seen something of the world.'

'A little—very little.'

'The South Seas, I believe. D'you know Papeete?'

'I have been there.'

'D'you know New Guinea at all?'

'No. I was never near it. I am better acquainted with the Far East—China, Japan, etc.'

Suddenly something, I shall never be able to tell what, prompted me to say:

'And the Andamans?'

The effect on my companion was as sudden as it was extraordinary. For a moment he staggered on the path like a drunken man; his face grew ashen pale, and he had to give utterance to a hoarse choking sound before be could get out a word. Then he said:

'No—no—you are quite mistaken, I assure you. I never knew the Andamans.'

Now, on the Andamans, as all the world knows, are located the Indian penal establishments, and noting his behaviour, I became more and more convinced in my own mind that there was some mystery about Mr Baxter that had yet to be explained. I had still a trump card to play.

'I'm afraid you are not very well, Mr Baxter,' I said at length. 'Perhaps the heat is too much for you, or we are walking too fast? This is my hotel. Won't you come inside and take a glass of wine or something to revive you?'

He nodded his head eagerly. Large drops of perspiration stood on his forehead, and I saw that he was quite unstrung.

'I am not well—not at all well.'

As soon as we reached the smoking-room I rang for two brandies and sodas. When they arrived he drank his off almost at a gulp, and in a few seconds was pretty well himself again.

'Thank you for your kindness, Mr Hatteras,' he said. 'I think we must have walked up the hill a little too fast for my strength. Now, I must be going back to the town. I find I have forgotten something.'

Almost by instinct I guessed his errand. He was going to dispatch another telegram. Resolved to try the effect of one parting shot, I said:

'Perhaps you do not happen to be going near the telegraph office again? If you are, should I be taxing your kindness too much if I asked you to leave a message there for me? I find *I* have forgotten one.'

He bowed and simply said:

'With much pleasure.'

He pronounced it 'pleesure', and as he said it he licked his lips in his usual self-satisfied fashion. I wondered how he would conduct himself when he saw the message I was going to write.

Taking a form from a table near where I sat, I wrote the following:

> John Nicholson,
>   Langham Hotel, London.
> The train is laid, but a new danger has arisen.
>   HATTERAS.

Blotting it carefully, I gave it into his hands, at the same time asking him to read it, lest my writing should not be decipherable and any question might be asked concerning it. As he read I watched his face intently. Never shall I forget the expression that swept over it. I had scored a complete victory. The shaft went home. But only for an instant. With wonderful alacrity he recovered himself and, shaking me feebly by the hand, bade me goodbye, promising to see that my message was properly delivered.

When he had gone I laid myself back in my chair for a good think. The situation was a peculiar one in every way. If he were up to some devilry I had probably warned him. If not, why had he betrayed himself so openly?

Half an hour later an answer to my first telegram arrived, and, such is the working of Fate, it necessitated my immediate return to London. I had been thinking of going for some days past, but had put it off. Now it was decided for me.

As I did not know whether I should return to Bournemouth again, I determined to call upon the Marquis to bid him goodbye. Accordingly, donning my hat, I set off for the house.

Now if Burke may be believed, the Duke of Glenbarth possesses houses in half the counties of the kingdom; but I am told his seaside residence takes precedence of them all in his affections. Standing well out on the cliffs, it commands a lovely view of the bay—looks toward the Purbeck Hills on the right, and the Isle of Wight and Hengistbury Head on the left. The house itself, as far as I could see, left nothing to be desired, and the grounds had been beautified in the highest form of landscape gardening.

I found my friend and his father in a summer-house upon the lawn. Both appeared unaffectedly glad to see me, and equally sorry to hear that I had come to bid them goodbye. Mr Baxter was not visible, and it was with no little surprise I learned that he, too, was contemplating a trip to the metropolis.

'I hope, if ever you visit Bournemouth again, you will come and see us,' said the Duke as I rose to leave.

'Thank you,' said I, 'and I hope if ever your son visits Australia you will permit me to be of some service to him.'

'You are very kind. I will bear your offer in mind.'

Shaking hands with them both, I bade them goodbye, and went out through the gate.

But I was not to escape without an interview with my clerical friend after all. As I left the grounds and turned into the public road I saw a man emerge from a little wicket gate some fifty yards or so further down the hedge. From the way he made his appearance, it was obvious he had been waiting for me to leave the house.

It was, certainly enough, my old friend Baxter. As I came up with him he said, with the same sanctimonious grin that usually encircled his mouth playing round it now:

'A nice evening for a stroll, Mr Hatteras.'

'A very nice evening, as you say, Mr Baxter.'

'May I intrude myself upon your privacy for five minutes?'

'With pleasure. What is your business?'

'Of small concern to you, sir, but of immense importance to me. Mr Hatteras, I have it in my mind that you do not like me.'

'I hope I have not given you cause to think so. Pray what can have put such a notion into your head?'

I half hoped that he would make some allusion to the telegram he had dispatched for me that morning, but he was far too cunning for that. He looked me over and over out of his small ferrety eyes before he replied:

'I cannot tell you why I think so, Mr Hatteras, but instinct generally makes us aware when we are not quite all we might be to other people. Forgive me for speaking in this way to you, but you must surely see how much it means to me to be on good terms with friends of my employer's family.'

'You are surely not afraid lest I should prejudice the Duke against you?'

'Not afraid, Mr Hatteras! I have too much faith in your sense of justice to believe that you would willingly deprive me of my means of livelihood—for of course that is what it would mean in plain English.'

'Then you need have no fear. I have just said goodbye to them. I am going away tomorrow, and it is very improbable that I shall ever see either of them again.'

'You are leaving for Australia?'

'Very shortly, I think.'

'I am much obliged to you for the generous way you have treated me. I shall never forget your kindness.'

'Pray don't mention it. Is that all you have to say to me? Then good-evening!'

'Good-evening, Mr Hatteras.'

He turned back by another gate into the garden, and I continued my way along the cliff, reflecting on the curious interview I had just passed through. If the truth must be known, I was quite at a loss to understand what he meant by it! Why had he asked that question about Australia? Was it only chance that had led him to put it, or was it done designedly, and for some reason connected with that mysterious 'train' mentioned in his telegram?

I was to find out later, and only too thoroughly!

CHAPTER VI

I meet Dr Nikola again

It is strange with what ease, rapidity, and apparent unconsciousness the average man jumps from crisis to crisis in that strange medley he is accustomed so flippantly to call His Life. It was so in my case. For two days after my return from Bournemouth I was completely immersed in the toils of Hatton Garden, had no thought above the sale of pearls and the fluctuations in the price of shell; yet, notwithstanding all this, the afternoon of the third day found me kicking my heels on the pavement of Trafalgar Square, my mind quite made up, my passage booked, and my ticket for Australia stowed away in my waistcoat pocket.

As I stood there the grim, stone, faces of the lions above me were somehow seen obscurely, Nelson's monument was equally unregarded, for my thoughts were far away with my mind's eye, following an ocean mail-steamer as she threaded her tortuous way between the Heads and along the placid waters of Sydney Harbour.

So wrapped up was I in the folds of this agreeable reverie, that when I felt a heavy hand upon my shoulder and heard a masculine voice say joyfully in my ear, 'Dick Hatteras, or I'm a Dutchman,' I started as if I had been shot.

Brief as was the time given me for reflection, it was long enough for that voice to conjure up a complete scene in my mind. The last time I had heard it was on the bridge of the steamer *Yarraman*, lying in the land-locked harbour of Cairns, on the Eastern Queensland coast; a canoeful of darkies were jabbering alongside, and a cargo of bananas was being shipped aboard.

I turned and held out my hand.

'Jim Percival!' I cried, with as much pleasure as astonishment. 'How on earth does it come about that you are here?'

'Arrived three days ago,' the good-looking young fellow replied. 'We're lying in the River just off the West India Docks. The old man kept us at it like galley slaves till I began to think we should never get the cargo out. Been up to the office this morning, coming back saw you standing here looking as if you were thinking of something ten thousand miles away. I tell you I nearly jumped out of my skin with astonishment, thought there couldn't be two men with the same face and build, so smacked you on the back, discovered I was right, and here we are. Now spin your yarn. But stay, let's first find a more convenient place than this.'

We strolled down the Strand together, and at last had the good fortune to discover a 'house of call' that met with even his critical approval. Here I narrated as much of my doings since we had last met, as I thought would satisfy his curiosity. My meeting with that mysterious individual at the French restaurant and my suspicions of Baxter particularly amused him.

'What a rum beggar you are, to be sure!' was his disconcerting criticism when I had finished. 'What earthly reason have you for thinking that this chap, Baxter, has any designs upon your young swell, Beckenham, or whatever his name may be?'

I meet Dr Nikola again

'What makes you stand by to shorten sail when you see a suspicious look about the sky? Instinct, isn't it?'

'That's a poor way out of the argument, to my thinking.'

'Well, at any rate, time will show how far I'm right or wrong; though I don't suppose I shall hear any more of the affair, as I return to Australia in the *Saratoga* on Friday next.'

'And what are you going to do now?'

'I haven't the remotest idea. My business is completed, and I'm just kicking my heels in idleness till Friday comes and it is time for me to set off for Plymouth.'

'Then I have it. You'll just come along down to the docks with me; I'm due back at the old hooker at five sharp. You'll dine with us— pot luck, of course. Your old friend Riley is still chief officer; I'm second; young Cleary, whom you remember as apprentice, is now third; and, if I'm not very much mistaken, we'll find old Donald Maclean aboard too, tinkering away at his beloved engines. I don't believe that fellow could take a holiday away from his thrust blocks and piston rods if he were paid to. We'll have a palaver about old times, and I'll put you ashore myself when you want to go. There, what do you say?'

'I'm your man,' said I, jumping at his offer with an alacrity which must have been flattering to him.

The truth was, I was delighted to have secured some sort of companionship, for London, despite its multitudinous places of amusement, and its five millions of inhabitants, is but a dismal caravanserai to be left alone in. Moreover, the *Yarraman*'s officers and I were old friends, and, if the truth must be told, my heart yearned for the sight of a ship and a talk about days gone by.

Accordingly, we made our way down to the Embankment, took the underground train at Charing Cross for Fenchurch Street, proceeding thence by 'The London and Blackwall' to the West India Docks.

The *Yarraman*, travel-stained, and bearing on her weather-beaten plates evidences of the continuous tramp-like life she had led, lay well out in the stream. Having chartered a waterman, we were put on board, and I had the satisfaction of renewing my acquaintance with the chief officer, Riley, at the yawning mouth of the for'ard hatch. The whilom apprentice, Cleary, now raised to the dignity of third officer, grinned a welcome to me from among the disordered raffle of the fo'c's'le head, while that excellent artificer, Maclean, oil-can and spanner in hand, greeted me affectionately in Gaelic from the entrance to the engine-room. The skipper was ashore, so I seated myself on the steps leading to the hurricane deck, and felt at home immediately.

Upon the circumstances attending that reunion there is no necessity for me to dwell. Suffice it that we dined in the deserted saloon, and adjourned later to my friend Percival's cabin in the alleyway just for'ard of the engine-room, where several bottles of Scotch whisky, a strange collection of glassware, and an assortment of excellent cigars, were produced. Percival and Cleary, being the juniors, ensconced themselves on the top bunk; Maclean (who had been induced to abandon his machinery in honour of our meeting) was given the washhandstand. Riley took the cushioned locker in the corner, while I, as their guest, was permitted the luxury of a canvas-backed deck-chair, the initials on the back of which were not those of its present owner. At first the conversation was circumscribed, and embraced Plimsoll, the attractions of London, and the decline in the price of freight; but, as the contents of the second bottle waned, speech became more unfettered, and the talk drifted into channels and latitudes widely different. Circumstances connected with bygone days were recalled; the faces of friends long hidden in the mists of time were brought again to mind; anecdotes illustrative of various types of maritime character succeeded each other in brisk succession, till Maclean, without warning, finding his voice, burst into incongruous melody. One song suggested another; a banjo was produced, and tuned to the noise of clinking glasses; and every moment the atmosphere grew thicker.

How long this concert would have lasted I cannot say, but I remember, after the third repetition of the chorus of the sea-chanty that might have been heard a mile away, glancing at my watch and discovering to my astonishment that it was past ten o'clock. Then rising to my feet I resisted all temptations to stay the night, and reminded my friend Percival of his promise to put me ashore again. He was true to his word, and five minutes later we were shoving off from the ship's side amid the valedictions of my hosts. I have a recollection to this day of the face of the chief engineer gazing sadly down upon me from the bulwarks, while his quavering voice asserted the fact, in dolorous tones, that

> Aft hae I rov'd by bonny Doon,
>   To see the rose and woodbine twine;
> And ilka bird sang o' its luve,
>   And fondly sae did I o' mine.

With this amorous farewell still ringing in my ears I landed at Limehouse Pier, and bidding my friend goodbye betook myself by the circuitous route of Emmett and Ropemaker Streets and Church Row to that aristocratic thoroughfare known as the East India Dock Road.

The night was dark and a thick rain was falling, presenting the mean-looking houses, muddy road, and foot-stained pavements in an aspect that was even more depressing than was usual to them. Despite the inclemency of the weather and the lateness of the hour, however, the street was crowded; blackguard men and foul-mouthed women, such a class as I had never in all my experience of rough folk encountered before, jostled each other on the pavements with scant ceremony; costermongers cried their wares, small boys dashed in and out of the crowd at top speed, and flaring gin palaces took in and threw out continuous streams of victims.

For some minutes I stood watching this melancholy picture, contrasting it with others in my mind. Then turning to my left hand I pursued my way in the direction I imagined the Stepney railway station to lie. It was not pleasant walking, but I was interested in the life about me—the people, the shops, the costermongers' barrows, and I might even say the public-houses.

I had not made my way more than a hundred yards along the street when an incident occurred that was destined to bring with it a train of highly important circumstances. As I crossed the entrance to a small side street, the door of an ill-looking tavern was suddenly thrust open and the body of a man was propelled from it, with a considerable amount of violence, directly into my arms. Having no desire to act as his support I pushed him from me, and as I did so glanced at the door through which he had come. Upon the glass was a picture, presumably nautical, and under it this legend 'The Green Sailor'. In a flash Bournemouth post office rose before my mind's eye, the startled face of Baxter on the doorstep, the swinging pencil on the telegraph stand, and the imprint of the mysterious message addressed to 'Nikola, Green Sailor Hotel, East India Dock Road'. So complete was my astonishment that at first I could do nothing but stand stupidly staring at it, then my curiosity asserted itself and, seeking the private entrance, I stepped inside. A short passage conducted me to a small and evil-smelling room abutting on the bar. On the popular side of the counter the place was crowded; in the chamber where I found myself I was the sole customer. A small table stood in the centre, and round this two or three chairs were ranged, while several pugnacious prints lent an air of decoration to the walls.

On the other side, to the left of that through which I had entered, a curtained doorway hinted at a similar room beyond. A small but heavily-built man, whom I rightly judged to be the landlord, was busily engaged with an assistant, dispensing liquor at the counter, but when I rapped upon the table he forsook his customers, and came to learn my wishes. I called for a glass of whisky, and seated myself at the table preparatory to commencing my enquiries as to the existence of Baxter's mysterious friend. But at the moment that I was putting my first question the door behind the half-drawn curtain, which must have been insecurely fastened, opened about an inch, and a voice greeted my ears that brought me up all standing with surprise. *It was the voice of Baxter himself.*

'I assure you,' he was saying, 'it was desperate work from beginning to end, and I was never so relieved in my life as when I discovered that he had really come to say goodbye.'

At this juncture one of them must have realized that the door was open, for I heard some one rise from his chair and come towards it. Acting under the influence of a curiosity, which was as baneful to himself as it was fortunate for me, before closing it he opened the door wider and looked into the room where I sat. It was Baxter, and if I live to be an hundred I shall not forget the expression on his face as his eyes fell upon me.

'Mr Hatteras!' he gasped, clutching at the wall for support.

Resolved to take him at a disadvantage, I rushed towards him and shook him warmly by the hand, at the same time noticing that he had discarded his clerical costume. It was too late now for him to pretend that he did not know me, and as I had taken the precaution to place my foot against it, it was equally impossible for him to shut the door. Seeing this he felt compelled to surrender, and I will do him the justice to admit that he did it with as good a grace as possible.

'Mr Baxter,' I said, 'this is the last place I should have expected to meet you in. May I come in and sit down?'

Without giving him time to reply I entered the room, resolved to see who his companion might be. Of course, in my own mind I had quite settled that it was the person to whom he had telegraphed from Bournemouth—in other words Nikola. But who was Nikola? And had I ever seen him before?

My curiosity was destined to be satisfied, and in a most unexpected fashion. For there, sitting at the table, a half-smoked cigarette between his fingers, and his face turned towards me, was the man whom I had seen playing chess in the restaurant, the man who had told me my name by the cards in my pocket, and the man who had warned me in such a mysterious fashion about my sweetheart's departure. He was Baxter's correspondent! He was Nikola!

Whatever my surprise may have been, he was not in the least disconcerted, but rose calmly from his seat and proffered me his hand, saying as he did so:

'Good-evening, Mr Hatteras. I am delighted to see you, and still more pleased to learn that you and my worthy old friend, Baxter, have met before. Won't you sit down?'

I seated myself on a chair at the further end of the table; Baxter meanwhile looked from one to the other of us as if uncertain whether to go or stay. Presently, however, he seemed to make up his mind, and advancing towards Nikola, said, with an earnestness that I could see was assumed for the purpose of putting me off the scent:

'And so I cannot induce you, Dr Nikola, to fit out an expedition for the work I have named?'

'If I had five thousand pounds to throw away,' replied Nikola, 'I might think of it, Mr Baxter, but as I haven't you must understand that it is impossible.' Then seeing that the other was anxious to be going, he continued, 'Must you be off? then good-night.'

Baxter shook hands with us both with laboured cordiality, and having done so slunk from the room. When the door closed upon him Nikola turned to me.

'There must be some fascination about a missionary's life after all,' he said. 'My old tutor, Baxter, as you are aware, has a comfortable position with the young Marquis of Beckenham, which, if he conducts himself properly, may lead to something really worth having in the future, and yet here he is anxious to surrender it in order to go back to his missionary work in New Guinea, to his hard life, insufficient food, and almost certain death.'

'He was in New Guinea then?'

'Five years—so he tells me.'

'Are you certain of that?'

'Absolutely!'

'Then all I can say is that, in spite of his cloth, Mr Baxter does not always tell the truth.'

'I am sorry you should think that. Pray what reason have you for saying so?'

'Simply because in a conversation I had with him at Bournemouth he deliberately informed me that he had never been near New Guinea in his life.'

'You must have misunderstood him. However that has nothing to do with us. Let us turn to a pleasanter subject.'

He rang the bell, and the landlord having answered it, ordered more refreshment. When it arrived he lit another cigarette, and leaning back in his chair glanced at me through half-closed eyes.

Then occurred one of the most curious and weird circumstances connected with this meeting. Hardly had he laid himself back in his chair before I heard a faint scratching against the table leg, and next moment an enormous cat, black as the Pit of Tophet, sprang with a bound upon the table and stood there steadfastly regarding me, its eyes flashing and its back arched. I have seen cats without number, Chinese, Persian, Manx, the Australian wild cat, and the English tabby, but never in the whole course of my existence such another as that owned by Dr Nikola. When it had regarded me with its evil eyes for nearly a minute, it stepped daintily across to its master, and rubbed itself backwards and forwards against his arm, then to my astonishment it clambered up on to his shoulder and again gave me the benefit of its fixed attention. Dr Nikola must have observed the amazement depicted in my face, for he smiled in a curious fashion, and coaxing the beast down into his lap fell to stroking its fur with his long, white fingers. It was as uncanny a performance as ever I had the privilege of witnessing.

'And so, Mr Hatteras,' he said slowly, 'you are thinking of leaving us?'

'I am,' I replied, with a little start of natural astonishment. 'But how did you know it?'

'After the conjuring tricks—we agreed to call them conjuring tricks, I think—I showed you a week or two ago, I wonder that you

I meet Dr Nikola again

should ask such a question. You have the ticket in your pocket even now.'

All the time he had been speaking his extraordinary eyes had never left my face; they seemed to be reading my very soul, and his cat ably seconded his efforts.

'By the way, I should like to ask you a few questions about those self-same conjuring tricks,' I said. 'Do you know you gave me a most peculiar warning?'

'I am very glad to hear it; I hope you profited by it.'

'It cost me a good deal of uneasiness, if that's any consolation to you. I want to know how you did it!'

'My fame as a wizard would soon evaporate if I revealed my methods,' he answered, still looking steadfastly at me. 'However, I will give you another exhibition of my powers, if you like. In fact, another warning. Have you confidence enough in me to accept it?'

'I'll wait and see what it is first,' I replied cautiously, trying to remove my eyes from his.

'Well, my warning to you is this—you intend to sail in the *Saratoga* for Australia on Friday next, don't you? Well, then, don't go; as you love your life, don't go!'

'Good gracious! and why on earth not?' I cried.

He stared fixedly at me for more than half a minute before he answered. There was no escaping those dreadful eyes, and the regular sweep of those long white fingers on the cat's black fur seemed to send a cold shiver right down my spine. Bit by bit I began to feel a curious sensation of dizziness creeping over me.

'Because you will *not* go. You cannot go. I forbid you to go.'

I roused myself with an effort, and sprang to my feet, crying as I did so:

'And what right have *you* to forbid me to do anything? I'll go on Friday, come what may. And I'd like to see the man who will prevent me.'

Though he must have realized that his attempt to hypnotize me (for attempt it certainly was) had proved a failure, he was not in the least disconcerted.

'My dear fellow,' he murmured gently, knocking off the ash of his cigarette against the table edge as he did so, 'no one is seeking to prevent you. I gave you, at your own request—you will do me the justice to admit that—a little piece of advice. If you do not care to follow it, that is your concern, not mine; but pray do not blame me. Must you really go now? Then good-night, and goodbye, for I don't suppose I shall see you this side of the Line again.'

I took his proffered hand, and wished him good-night. Having done so, I left the house, heartily glad to have said goodbye to the only man in my life whom I have really feared.

When in the train, on my way back to town, I came to review the meeting in the 'Green Sailor', I found myself face to face with a series of problems very difficult to work out. How had Nikola first learned my name? How had he heard of the Wetherells? Was he the mysterious person his meeting with whom had driven Wetherell out of England? Why had Baxter telegraphed to him that 'the train was laid'? Was I the new danger that had arisen? How had Baxter come to be at the 'Green Sailor' in non-clerical costume? Why had he been so disturbed at my entry? Why had Nikola invented such a lame excuse to account for his presence there? Why had he warned me not to sail in the *Saratoga*? and, above all, why had he resorted to hypnotism to secure his ends?

I asked myself these questions, but one by one I failed to answer them to my satisfaction. Whatever other conclusion I might have come to, however, one thing at least was certain: that was, that my original supposition was a correct one. There was a tremendous mystery somewhere. Whether or not I was to lose my interest in it after Friday remained to be seen.

Arriving at Fenchurch Street, I again took the Underground, and, bringing up at the Temple, walked to my hotel off the Strand. It was nearly twelve o'clock by the time I entered the hall; but late as it was I found time to examine the letter rack. It contained two envelopes bearing my name, and taking them out I carried them with me to my room. One, to my delight, bore the postmark of Port Said, and was addressed in my sweetheart's handwriting. You may guess how eagerly I tore it open, and with what avidity I devoured its contents. From it I gathered that they had arrived at the entrance of the Suez Canal safely; that her father had recovered his spirits more and more with every mile that separated him from Europe. He was now almost himself again, she said, but still refused with characteristic determination to entertain the smallest notion of myself as a son-in-law. But Phyllis herself did not despair of being able to talk him round. Then came a paragraph which struck me as being so peculiar as to warrant my reproducing it here:

> The passengers, what we have seen of them, appear to be, with one exception, a nice enough set of people. That exception, however, is intolerable; his name is Prendergast, and his personal appearance is as objectionable as his behaviour is extraordinary; his hair is snow-white, and his face is deeply pitted with smallpox. This is, of course, not his fault, but it seems somehow to aggravate the distaste I have for him. Unfortunately we were thrown into his company in Naples, and since then the creature has so far presumed upon that introduction, that he scarcely leaves me alone for a moment. Papa does not seem to mind him so much, but I continually thank goodness that, as he leaves the boat in Port Said, the rest of the voyage will be performed without him.

The remainder of the letter has no concern for anyone but myself, so I do not give it. Having read it I folded it up and put it in my pocket, feeling that if I had been on board the boat I should in all probability have allowed Mr Prendergast to understand that his attentions were distasteful and not in the least required. If I could only have foreseen that within a fortnight I was to be enjoying the doubtful pleasure of that very gentleman's society, under circumstances as important as life and death, I don't doubt I should have thought still more strongly on the subject.

The handwriting of the second envelope was bold, full of character, but quite unknown to me. I opened it with a little feeling of curiosity, and glanced at the signature, 'Beckenham'.

It ran as follows:

West Cliff, Bournemouth,
Tuesday Evening.

MY DEAR MR HATTERAS,

I have great and wonderful news to tell you! This week has proved an extraordinarily eventful one for me, for what do you think? My father has suddenly decided that I shall travel. All the details have been settled in a great hurry. You will understand this when I tell you that Mr Baxter and I sail for Sydney in the steamship *Saratoga* next week. My father telegraphed to Mr Baxter, who is in London, to book our passages and to choose our cabins this morning. I can only say that my greatest wish is that you were coming with us. Is it so impossible? Cannot you make your arrangements fit in? We shall travel overland to Naples and join the boat there. This is Mr Baxter's proposition, and you may be sure, considering what I shall see *en route*, I have no objection to urge against it. Our tour will be an extensive one. We visit Australia and New Zealand, go thence to Honolulu, thence to San Francisco, returning, across the United States, via Canada, to Liverpool.

You may imagine how excited I am at the prospect, and as I feel that I owe a great measure of my good fortune to you, I want to be the first to acquaint you of it.

Yours ever sincerely,
BECKENHAM.

I read the letter through a second time, and then sat down on my bed to think it out. One thing was self-evident. I knew now how Nikola had become aware that I was going to sail in the mail boat on Friday; Baxter had seen my name in the passenger list, and had informed him.

I undressed and went to bed, but not to sleep. I had a problem to work out, and a more than usually difficult one it was. Here was the young Marquis of Beckenham, I told myself, only son of his father, heir to a great name and enormous estates, induced to travel by my representations. There was a conspiracy afoot in which, I could not help feeling certain, the young man was in some way involved. And yet I had no right to be certain about it after all, for my suspicions at best were only conjectures. Now the question was whether I ought to warn the Duke or not? If I did I might be frightening him without cause, and might stop his son's journey; and if I did not, and things went wrong—well, in that case, I should be the innocent means of bringing a great and lasting sorrow upon his house. Hour after hour I turned this question over and over in my mind, uncertain how to act. The clocks chimed their monotonous round, the noises died down and rose again in the streets, and daylight found me only just come to a decision. I would *not* tell him; but at the same time I would make doubly sure that I sailed aboard that ship myself, and that throughout the voyage I was by the young man's side to guard him from ill.

Breakfast time came, and I rose from my bed wearied with thought. Even a bath failed to restore my spirits. I went downstairs and, crossing the hall again, examined the rack. Another letter awaited me. I passed into the dining-room and, seating myself at my table, ordered breakfast. Having done so, I turned to my correspondence. Fate seemed to pursue me. On this occasion the letter was from the lad's father, the Duke of Glenbarth himself, and ran as follows:

I meet Dr Nikola again

Sandridge Castle, Bournemouth,
Wednesday.

DEAR MR HATTERAS,

My son tells me he has acquainted you with the news of his departure for Australia next week. I don't doubt this will cause you some little surprise; but it has been brought about by a curious combination of circumstances. Two days ago I received a letter from my old friend, the Earl of Amberley, who, as you know, has for the past few years been Governor of the colony of New South Wales, telling me that his term of office will expire in four months. Though he has not seen my boy since the latter was two years old, I am anxious that he should be at the head of affairs when he visits the colony. Hence this haste. I should have liked nothing better than to have accompanied him myself, but business of the utmost importance detains me in England. I am, however, sending Mr Baxter with him, with powerful credentials, and if it should be in your power to do anything to assist them you will be adding materially to the debt of gratitude I already owe you.

Believe me, my dear Mr Hatteras, to be,
Very truly yours,
GLENBARTH.

My breakfast finished, I answered both these letters, informed my friends of my contemplated departure by the same steamer, and promised that I would do all that lay in my power to ensure both the young traveller's pleasure and his safety.

For the rest of the morning I was occupied inditing a letter to my sweetheart, informing her of my return to the Colonies, and telling her all my adventures since her departure.

The afternoon was spent in saying goodbye to the few business friends I had made in London, and in the evening I went for the last time to a theatre.

Five minutes to eleven o'clock next morning found me at Waterloo sitting in a first-class compartment of the West of England express, bound for Plymouth and Australia. Though the platform was crowded to excess, I had the carriage so far to myself, and was about to congratulate myself on my good fortune, when a porter appeared on the scene, and deposited a bag in the opposite corner. A moment later, and just as the train was in motion, a man jumped in the carriage, tipped the servant, and then placed a basket upon the rack. The train was half-way out of the station before he turned round, and my suspicions were confirmed. *It was Dr Nikola!*

Though he must have known who his companion was, he affected great surprise.

'Mr Hatteras,' he cried, 'I think this is the most extraordinary coincidence I have ever experienced in my life.'

'Why so?' I asked. 'You knew I was going to Plymouth today, and one moment's reflection must have told you, that as my boat sails at eight, I would be certain to take the morning express, which lands me there at five. Should I be indiscreet if I asked where you may be going?'

'Like yourself, I am also visiting Plymouth,' he answered, taking the basket, before mentioned, down from the rack, and drawing a French novel from his coat pocket. 'I expect an old Indian friend home by the mail boat that arrives tonight. I am going down to meet him.'

I felt relieved to hear that he was not thinking of sailing in the *Saratoga*, and after a few polite commonplaces, we both lapsed into silence. I was too suspicious, and he was too wary, to appear over friendly. Clapham, Wimbledon, Surbiton, came and went. Weybridge and Working flashed by at lightning speed, and even Basingstoke was reached before we spoke again. That station behind us, Dr Nikola took the basket before mentioned on his knee, and opened it. When he had done so, the same enormous black cat, whose acquaintance I had made in the East India Dock Road, stepped proudly forth. In the daylight the brute looked even larger and certainly fiercer than before. I felt I should have liked nothing better than to have taken it by the tail and hurled it out of the window. Nikola, on the other hand, seemed to entertain for it the most extraordinary affection.

Now such was this marvellous man's power of fascination that by the time we reached Andover Junction his conversation had roused me quite out of myself, had made me forget my previous distrust of him, and enabled me to tell myself that this railway journey was one of the most enjoyable I had ever undertaken.

In Salisbury we took luncheon baskets on board, with two bottles of champagne, for which my companion, in spite of my vigorous protest, would insist upon paying.

As the train rolled along the charming valley, in which lie the miniature towns of Wilton, Dinton, and Tisbury, we pledged each other in right good fellowship, and by the time Exeter was reached were friendly enough to have journeyed round the world together.

I meet Dr Nikola again

Exeter behind us, I began to feel drowsy, and before the engine came to a standstill at Okehampton was fast asleep.

I remember no more of that ill-fated journey; nor, indeed have I any recollection of anything at all, until I woke up in Room No. 37 of the Ship and Vulture Hotel in Plymouth.

The sunshine was streaming in through the slats of the Venetian blinds, and a portly gentleman, with a rosy face, and grey hair, was standing by my bedside, holding my wrist in his hand, and calmly scrutinizing me. A nurse in hospital dress stood beside him.

'I think he'll do now,' he said to her as he rubbed his plump hands together; 'but I'll look round in the course of the afternoon.'

'One moment,' I said feebly, for I found I was too weak to speak above a whisper. 'Would you mind telling me where I am, and what is the matter with me?'

'I should very much like to be able to do so,' was the doctor's reply. 'My opinion is, if you want me to be candid, that you have been drugged and well-nigh poisoned, by a remarkably clever chemist. But what the drug and the poison were, and who administered it to you, and the motive for doing so, is more than I can tell you. From what I can learn from the hotel proprietors you were brought here from the railway station in a cab last night by a gentleman who happened to find you in the carriage in which you travelled down from London. You were in such a curious condition that I was sent for and this nurse procured. Now you know all about it.'

'What day did you say this is?'

'Saturday, to be sure.'

'Saturday!' I cried. 'You don't mean that! Then, by Jove, I've missed the *Saratoga* after all. Here, let me get up! And tell them downstairs to send for the Inspector of Police. I have got to get to the bottom of this.'

I sat up in bed, but was only too glad to lie down again, for my weakness was extraordinary. I looked at the doctor.

'How long before you can have me fit to travel?'

'Give yourself three days' rest and quiet,' he replied, 'and we'll see what we can do.'

'Three days? And two days and a half to cross the continent, that's five and a half—say six days. Good! I'll catch the boat in Naples, and then, Dr Nikola, if you're aboard, as I suspect, I should advise you to look out.'

CHAPTER VII

Port Said, and what Befell us There

Fortunately for me my arrangements fitted in exactly, so that at one thirty p.m., on the seventh day after my fatal meeting with Dr Nikola in the West of England express, I had crossed the continent, and stood looking out on the blue waters of Naples Bay. To my right was the hill of San Martino, behind me that of Capo di Monte, while in the distance, to the southward, rose the cloud-tipped summit of Vesuvius.

The journey from London is generally considered, I believe, a long and wearisome one; it certainly proved so to me, for it must be remembered that my mind was impatient of every delay, while my bodily health was not as yet recovered from the severe strain that had been put upon it.

The first thing to be done on arrival at the terminus was to discover a quiet hotel; a place where I could rest and recoup during the heat of the day, and, what was perhaps more important, where I should run no risk of meeting with Dr Nikola or his satellites. I had originally intended calling at the office of the steamship company in order to explain the reason of my not joining the boat in Plymouth, planning afterwards to cast about me, among the various hotels, for the Marquis of Beckenham and Mr Baxter. But, on second thoughts, I saw the wisdom of abandoning both these courses. If you have followed the thread of my narrative, you will readily understand why.

Nor for the same reason did I feel inclined to board the steamer, which I could see lying out in the harbour, until darkness had fallen. I ascertained, however, that she was due to sail at midnight, and that the mails were already being got aboard.

Almost exactly as eight o'clock was striking, I mounted the gangway, and strolled down the promenade deck to the first saloon entrance; then calling a steward to my assistance, I had my baggage conveyed to my cabin, where I set to work arranging my little knicknacks, and making myself comfortable for the five weeks' voyage that lay before us. So far I had seen nothing of my friends, and, on making enquiries, I discovered that they had not yet come aboard.

Indeed, they did not do so until the last boat had discharged its burden at the gangway. Then I met Lord Beckenham on the promenade deck, and unaffected was the young man's delight at seeing me.

'Mr Hatteras,' he cried, running forward to greet me with outstretched hand, 'this was all that was wanting to make my happiness complete. I *am* glad to see you. I hope your cabin is near ours.'

'I'm on the port side just abaft the pantry,' I answered, shaking him by the hand. 'But tell me about yourself. I expect you had a pleasant journey across the continent.'

'Delightful!' was his reply. 'We stayed a day in Paris, and another in Rome, and since we have been here we have been rushing about seeing everything, like a regulation pair of British tourists.'

At this moment Mr Baxter, who had been looking after the luggage, I suppose, made his appearance, and greeted me with more cordiality than I had expected him to show. To my intense surprise, however, he allowed no sign of astonishment to escape him at my having joined the boat after all. But a few minutes later, as we were approaching the companion steps, he said:

'I understood from his lordship, Mr Hatteras, that you were to embark at Plymouth; was I mistaken, therefore, when I thought I saw you coming off with your luggage this evening?'

'No, you were not mistaken,' I answered, being able now to account for his lack of surprise. 'I came across the continent like yourselves, and only joined the vessel a couple of hours ago.'

Here the Marquis chimed in, and diverted the conversation into another channel.

'Where is everybody?' he asked, when Mr Baxter had left us and gone below. 'There are a lot of names on the passenger list, and yet I see nobody about!'

'They are all in bed,' I answered. 'It is getting late, you see, and, if I am not mistaken, we shall be under way in a few minutes.'

'Then, I think, if you'll excuse me for a few moments, I'll go below to my cabin. I expect Mr Baxter will be wondering where I am.'

When he had left me I turned to the bulwarks and stood looking across the water at the gleaming lights ashore. One by one the boats alongside pushed off, and from the sounds that came from for'ard, I gathered that the anchor was being got aboard. Five minutes later we had swung round to our course and were facing for the open sea. For the first mile or so my thoughts chased each other in rapid succession. You must remember that it was in Naples I had learnt that my darling loved me, and it was in Naples now that I was bidding goodbye to Europe and to all the strange events that had befallen me there. I leant upon the rail, looked at the fast receding country in our wake, at old Vesuvius, fire-capped, away to port, at the Great Bear swinging in the heavens to the nor'ard, and then thought of the Southern Cross which, before many weeks were passed, would be lifting its head above our bows to welcome me back to the sunny land and to the girl I loved so well. Somehow I felt glad that the trip to England was over, and that I was really on my way home at last.

The steamer ploughed her almost silent course, and three-quarters of an hour later we were abreast of Capri. As I was looking at it, Lord Beckenham came down the deck and stood beside me. His first speech told me that he was still under the influence of his excitement; indeed, he spoke in rapturous terms of the enjoyment he expected to derive from his tour.

'But are you sure you will be a good sailor?' I asked.

'Oh, I have no fear of that,' he answered confidently. 'As you know, I have been out in my boat in some pretty rough weather and never felt in the least ill, so I don't think it is likely that I shall begin to be a bad sailor on a vessel the size of the *Saratoga*. By the way, when are we due to reach Port Said?'

'Next Thursday afternoon, I believe, if all goes well.'

'Will you let me go ashore with you if you go? I don't want to bother you, but after all you have told me about the place, I should like to see it in your company.'

'I'll take you with pleasure,' I answered, 'provided Mr Baxter gives his consent. I suppose we must regard him as skipper.'

'Oh, I don't think we need fear his refusing. He is very good-natured, you know, and lets me have my own way a good deal.'

'Where is he now?'

'Down below, asleep. He has had a lot of running about today and thought he would turn in before we got under way. I think I had better be going now. Good-night.'

'Good-night,' I answered, and he left me again.

When I was alone I returned to my thoughts of Phyllis and the future, and as soon as my pipe was finished, went below to my bunk. My berth mate I had discovered earlier in the evening was a portly

Port Said, and what Befell us There 89

English merchant of the old school, who was visiting his agents in Australia; and, from the violence of his snores, I should judge had not much trouble on his mind. Fortunately mine was the lower bunk, and, when I had undressed, I turned into it to sleep like a top until roused by the bathroom steward at half-past seven next morning. After a good bathe I went back to my cabin and set to work to dress. My companion by this time was awake, but evidently not much inclined for conversation. His usual jovial face, it struck me, was not as rosy as when I had made his acquaintance the night before, and from certain signs I judged that his good spirits were more than half assumed.

All this time a smart sea was running, and, I must own, the *Saratoga* was rolling abominably.

'A very good morning to you, my dear sir,' my cabin mate said, with an air of enjoyment his pallid face belied, as I entered the berth. 'Pray how do you feel today?'

'In first-class form,' I replied, 'and as hungry as a hunter.'

He laid himself back on his pillow with a remark that sounded very much like 'Oh dear,' and thereafter I was suffered to shave and complete my toilet in silence. Having done so I put on my cap and went on deck.

It was indeed a glorious morning; bright sunshine streamed upon the decks, the sea was a perfect blue, and so clear was the air that, miles distant though it was from us, the Italian coastline could be plainly discerned above the port bulwarks. By this time I had cross-examined the chief steward, and satisfied myself that Nikola was not aboard. His absence puzzled me considerably. Was it possible that I could have been mistaken in the whole affair, and that Baxter's motives were honest after all? But in that case why had Nikola drugged me? And why had he warned me against sailing in the *Saratoga*? The better to think it out I set myself for a vigorous tramp round the hurricane deck, and was still revolving the matter in my mind, when, on turning the corner by the smoking-room entrance, I found myself face to face with Baxter himself. As soon as he saw me, he came smiling towards me, holding out his hand.

'Good-morning, Mr Hatteras,' he said briskly; 'what a delightful morning it is, to be sure. You cannot tell how much I am enjoying it. The sea air seems to have made a new man of me already.'

'I am glad to hear it. And pray how is your charge?' I asked, more puzzled than ever by this display of affability.

'Not at all well, I am sorry to say.'

'Not well? You don't surely mean to say that he is seasick?'

'I'm sorry to say I do. He was perfectly well until he got out of his bunk half an hour ago. Then a sudden, but violent, fit of nausea seized him, and drove him back to bed again.'

'I am very sorry to hear it, I hope he will be better soon. He would have been one of the last men I should have expected to be bowled over. Are you coming for a turn round?'

'I shall feel honoured,' he answered, and thereupon we set off, step for step, for a constitutional round the deck. By the time we had finished it was nine o'clock, and the saloon gong had sounded for breakfast.

The meal over, I repaired to the Marquis's cabin, and having knocked, was bidden enter. I found my lord in bed, retching violently; his complexion was the colour of zinc, his hands were cold and clammy, and after every spasm his face streamed with perspiration.

'I am indeed sorry to see you like this,' I said, bending over him. 'How do you feel now?'

'Very bad indeed!' he answered, with a groan. 'I cannot understand it at all. Before I got out of bed this morning I felt as well as possible. Then Mr Baxter was kind enough to bring me a cup of coffee, and within five minutes of drinking it, I was obliged to go back to bed feeling hopelessly sick and miserable.'

'Well, you must try and get round as soon as you can, and come on deck; there's a splendid breeze blowing, and you'll find that will clear the sickness out of you before you know where you are.'

But his only reply was another awful fit of sickness, that made as if it would tear his chest asunder. While he was under the influence of it, his tutor entered, and set about ministering to him with a care and fatherly tenderness that even deceived me. I can see things more plainly now, on looking back at them, than I could then, but I must own that Baxter's behaviour towards the boy that morning was of a kind that would have hoodwinked the very Master of All Lies himself. I could easily understand now how this man had come to have such an influence over the kindly-natured Duke of Glenbarth, who, when all was said and done, could have had but small experience of men of Baxter's type.

Seeing that, instead of helping, I was only in the way, I expressed a hope that the patient would soon be himself again, and returned to the deck.

Luncheon came, and still Lord Beckenham was unable to leave his berth. In the evening he was no better. The following morning he was, if anything, stronger; but towards midday, just as he was thinking of getting up, his nausea returned upon him, and he was obliged to postpone the attempt. On Wednesday there was no improvement, and, indeed, it was not until Thursday afternoon, when the low-lying coast of Port Said was showing above the sea-line, that he felt in any way fit to leave his bunk. In all my experience of seasickness I had never known a more extraordinary case.

It was almost dark before we dropped our anchor off the town, and as soon as we were at a standstill I went below to my friend's cabin. He was sitting on the locker fully dressed.

'Port Said,' I announced. 'Now, how do you feel about going ashore? Personally, I don't think you had better try it.'

'Oh! but I want to go. I have been looking forward to it so much. I am much stronger than I was, believe me, and Mr Baxter doesn't think it could possibly hurt me.'

'If you don't tire yourself too much,' that gentleman put in.

'Very well, then,' I said. 'In that case I'm your man. There are plenty of boats alongside, so we'll have no difficulty about getting there. Won't you come, too, Mr Baxter?'

'I think not, thank you,' he answered. 'Port Said is not a place of which I am very fond, and as we shall not have much time here, I am anxious to utilize our stay in writing His Grace a letter detailing our progress so far.'

'In that case I think we had better be going,' I said, turning to his lordship.

We made our way on deck, and, after a little chaffering, secured a boat, in which we were pulled ashore. Having arrived there, we were immediately beset by the usual crowd of beggars and donkey boys, but withstanding their importunities, we turned into the Rue de Commerce and made our way inland. To my companion the crowded streets, the diversity of nationalities and costume, and the strange variety of shops and wares, were matters of absorbing interest. This will be the better understood when it is remembered that, poor though Port Said is in orientalism, it was nevertheless the first Eastern port he had encountered. We had both a few purchases to make, and this business satisfactorily accomplished, we hired a guide and started off to see the sights.

Passing out of the Rue de Commerce, our attention was attracted by a lame young beggar who, leaning on his crutches, blocked our way while he recited his dismal catalogue of woes. Our guide bade him be off, and indeed I was not sorry to be rid of him, but I could see, by glancing at his face, that my companion had taken his case more seriously. In fact we had not proceeded more than twenty yards before he asked me to wait a moment for him, and taking to his heels ran back to the spot where we had left him. When he rejoined us I said:

'You don't mean to say that you gave that rascal money?'

'Only half a sovereign,' he answered. 'Perhaps you didn't hear the pitiful story he told us? His father is dead, and now, if it were not for his begging, his mother and five young sisters would all be starving.'

I asked our guide if he knew the man, and whether his tale were true.

'No, monsieur,' he replied promptly, 'it is all one big lie. His father is in the jail, and, if she had her rights, his mother would be there too.'

Not another word was said on the subject, but I could see that the boy's generous heart had been hurt. How little he guessed the effect that outburst of generosity was to have upon us later on.

At our guide's suggestion, we passed from the commercial, through the European quarter, to a large mosque situated in Arab Town. It was a long walk, but we were promised that we should see something there that would amply compensate us for any trouble we might be put to to reach it. This turned out to be the case, but hardly in the fashion he had predicted.

The mosque was certainly a fine building, and at the time of our visit was thronged with worshippers. They knelt in two long lines, reaching from end to end, their feet were bare, and their heads turned towards the east. By our guide's instructions we removed our boots at the entrance, but fortunately, seeing what was to transpire later, took the precaution of carrying them into the building with us. From the main hall we passed into a smaller one, where a number of Egyptian standards, relics of the war of '82, were unrolled for our inspection. While we were examining them, our guide, who had for a moment left us, returned with a scared face to inform us that there were a number of English tourists in the mosque who had refused to take their boots off, and were evidently bent on making trouble. As he spoke the ominous hum of angry voices drifted in to us, increasing in volume as we listened. Our guide pricked up his ears and looked anxiously at the door.

'There will be trouble directly,' he said solemnly, 'if those young men do not behave themselves. If messieurs will be guided by me, they will be going. I can show them a backway out.'

For a moment I felt inclined to follow his advice, but Beckenham's next speech decided me to stay.

'You will not go away and leave those stupid fellows to be killed?' he said, moving towards the door into the mosque proper. 'However foolish they may have been, they are still our countrymen, and whatever happens we ought to stand by them.'

'If you think so, of course we will,' I answered, 'but remember it may cost us our lives. You still want to stay? Very good, then, come along, but stick close to me.'

We left the small ante-room, in which we had been examining the flags, and passed back into the main hall. Here an extraordinary scene presented itself.

In the furthest corner, completely hemmed in by a crowd of furious Arabs, were three young Englishmen, whose faces plainly showed how well they understood the dangerous position into which their own impudence and folly had enticed them.

Elbowing our way through the crowd, we reached their side, and immediately called upon them to push their way towards the big doors; but before this manœuvre could be executed, someone had given an order in Arabic, and we were all borne back against the wall.

'There is no help for it?' I cried to the biggest of the strangers. 'We must fight our way out. Choose your men and come along.'

So saying, I gave the man nearest me one under the jaw to remember me by, which laid him on his back, and then, having room to use my arms, sent down another to keep him company. All this time my companions were not idle, and to my surprise I saw the young Marquis laying about him with a science that I had to own afterwards did credit to his education. Our assailants evidently did not expect to meet with this resistance, for they gave way and began to back towards the door. One or two of them drew knives, but the space was too cramped for them to do much harm with them.

'One more rush,' I cried, 'and we'll turn them out.'

We made the rush, and next moment the doors were closed and barred on the last of them. This done, we paused to consider our position. True we had driven the enemy from the citadel, but then, unless we could find a means of escape, we ourselves were equally prisoners in it. What was to be done?

Leaving three of our party to guard the doors, the remainder searched the adjoining rooms for a means of escape; but though we were unsuccessful in our attempt to find an exit, we did what was the next best thing to it, discovered our cowardly guide in a corner, skulking in a curious sort of cupboard.

By the time we had proved to him that the enemy were really driven out, and that we had possession of the mosque, he recovered his wits a little, and managed, after hearing our promise to throw him to the mob outside unless he discovered a means of escape for us, to cudgel his brains and announce that he knew of one.

No sooner did we hear this, than we resolved to profit by it. The mob outside was growing every moment more impatient, and from the clang of steel-shod rifle butts on the stone steps we came to the conclusion that the services of a force of soldiery had been called in. The situation was critical, and twice imperious demands were made upon us to open the door. But, as may be supposed, this we did not feel inclined to do.

'Now, for your way out,' I said, taking our trembling guide, whose face seemed to blanch whiter and whiter with every knock upon the door, by the shoulders, and giving him a preliminary shake. 'Mind what you're about, and remember, if you lead us into any trap, I'll wring your miserable neck, as sure as you're alive. Go ahead.'

Collecting our boots and shoes, which, throughout the tumult, had been lying scattered about upon the floor, we passed into the anteroom, and put them on. Then creeping softly out by another door, we reached a small courtyard in the rear, surrounded on all sides by high walls. Our way, so our guide informed us, lay over one of these. But how we were to surmount them was a puzzle, for the lowest scaling place was at least twelve feet high. However, the business had to be done, and, what was more to the point, done quickly.

Calling the strongest of the tourists, who were by this time all quite sober, to my side, I bade him stoop down as if he were playing leapfrog; then, mounting his back myself, I stood upright, and stretched my arms above my head. To my delight my fingers reached to within a few inches of the top of the wall.

'Stand as steady as you can,' I whispered, 'for I'm going to jump.'

I did so, and clutched the edge. Now, if anybody thinks it is an easy thing to pull oneself to the top of the wall in that fashion, let him try it, and I fancy he'll discover his mistake. I only know I found it a harder business than I had anticipated, so much harder that when I reached the top I was so completely exhausted as to be unable to do anything for more than a minute. Then I whispered to another man to climb upon the first man's back, and stretch his hands up to mine. He did so, and I pulled him up beside me. The guide came next, then the other tourist, then Lord Beckenham. After which I took off and lowered my coat to the man who had stood for us all, and having done so, took a firm grip of the wall with my legs, and dragged him up as I had done the others.

It had been a longer business than I liked, and every moment, while we were about it, I expected to hear the cries of the mob inside the mosque, and to find them pouring into the yard to prevent our escape. The bolts on the door, however, must have possessed greater strength than we gave them credit for. At any rate, they did not give way.

When we were all safely on the wall, I asked the guide in which direction we should now proceed; he pointed to the adjoining roofs, and in Indian file, and with the stealthiness of cats, we accordingly crept across them.

The third house surmounted, we found ourselves overlooking a narrow alley, into which we first peered carefully, and, having discovered that no one was about, eventually dropped.

'Now,' said the guide, as soon as we were down, 'we must run along here, and turn to the left.'

We did so, to find ourselves in a broader street, which eventually brought us out into the thoroughfare through which we had passed to reach the mosque.

Having got our bearings now, we headed for the harbour, or at least for that part of the town with which I was best acquainted, as fast as our legs would carry us. But, startling as they had been, we had not yet done with adventures for the night.

Once in the security of the gaslit streets, we said goodbye to the men who had got us into all the trouble, and having come to terms with our guide, packed him off and proceeded upon our way alone.

Five minutes later the streaming lights of an open doorway brought us to a standstill, and one glance told us we were looking into the Casino. The noise of the roulette tables greeted our ears, and as we had still plenty of time, and my companion was not tired, I thought it a good opportunity to show him another phase of the seamy side of life.

But before I say anything about that I must chronicle a curious circumstance. As we were entering the building, something made me look round. To my intense astonishment I saw, or believed I saw, Dr Nikola standing in the street, regarding me. Bidding my companion remain where he was for a moment, I dashed out again and ran towards the place where I had seen the figure. But I was too late. If it were Dr Nikola, he had vanished as suddenly as he had come. I hunted here, there, and everywhere, in doorways, under verandahs, and down lanes, but it was no use, not a trace of him could I discover. So abandoning my search, I returned to the Casino. Beckenham was waiting for me, and together we entered the building.

The room was packed, and consequently all the tables were crowded, but as we did not intend playing, this was a matter of small concern to us. We were more interested in the players than the game. And, indeed, the expressions on the faces around us were extraordinary. On some hope still was in the ascendant, on others a haggard despair seemed to have laid its grisly hand; on everyone was imprinted the lust of gain. The effect on the young man by my side was peculiar. He looked from face to face, as if he were observing the peculiarities of some strange animals. I watched him, and then I saw his expression suddenly change.

Following the direction of his eyes, I observed a young man putting down his stake upon the board. His face was hidden from me, but by taking a step to the right I could command it. It was none other than the young cripple who had represented his parents to be in such poverty-stricken circumstances; the same young man whom Beckenham had assisted so generously only two hours before. As we looked, he staked his last coin, and that being lost, turned to leave the building. To do this, it was necessary that he should pass close by where we stood. Then his eyes met those of his benefactor, and with a look of what might almost have been shame upon his face, he slunk down the steps and from the building.

'Come, let us get out of this place,' cried my companion impatiently, 'I believe I should go mad if I stayed here long.'

Thereupon we passed out into the street, and without further ado proceeded in the direction in which I imagined the *Saratoga* to lie. A youth of about eighteen chequered summers requested, in broken English, to be permitted the honour of piloting us, but feeling confident of being able to find my way I declined his services.

For fully a quarter of an hour we plodded on, until I began to wonder why the harbour did not heave in sight. It was a queer part of the town we found ourselves in; the houses were perceptibly meaner and the streets narrower. At last I felt bound to confess that I was out of my reckoning, and did not know where we were.

'What are we to do?' asked my lord, looking at his watch. 'It's twenty minutes to eleven, and I promised Mr Baxter I would not be later than the hour.'

'What an idiot I was not to take that guide!'

The words were hardly out of my mouth before that personage appeared round the corner and came towards us. I hailed his coming with too much delight to notice the expression of malignant satisfaction on his face, and gave him the name of the vessel we desired to find. He appeared to understand, and the next moment we were marching off under his guidance in an exactly contrary direction.

We must have walked for at least ten minutes without speaking a word. The streets were still small and ill-favoured, but, as this was probably a short cut to the harbour, such minor drawbacks were not worth considering.

From one small and dirty street we turned into another and broader one. By this time not a soul was to be seen, only a vagrant dog or two lying asleep in the road. In this portion of the town gas lamps were at a discount, consequently more than half the streets lay in deep shadow. Our guide walked ahead, we followed half-a-dozen paces or so behind him. I remember noticing a Greek cognomen upon a signboard, and recalling a similar name in Thursday Island, when something very much resembling a thin cord touched my nose and fell over my chin. Before I could put my hand up to it it had begun to tighten round my throat. Just at the same moment I heard my companion utter a sharp cry, and after that I remember no more.

CHAPTER VIII

Our Imprisonment and Attempt at Escape

For what length of time I lay unconscious after hearing Beckenham's cry, and feeling the cord tighten round my throat, as narrated in the preceding chapter, I have not the remotest idea; I only know that when my senses returned to me again I found myself in complete darkness. The cord was gone from my neck, it is true, but something was still encircling it in a highly unpleasant fashion. On putting my hand up to it, to my intense astonishment, I discovered it to be a collar of iron, padlocked at the side, and communicating with a wall at the back by means of a stout chain fixed in a ring, which again was attached to a swivel.

This ominous discovery set me hunting about to find out where I was, and for a clue as to what these things might mean. That I was in a room was evident from the fact that, by putting my hands behind me, I could touch two walls forming a corner. But in what part of the town such a room might be was beyond my telling. One thing was evident, however, the walls were of brick, unplastered and quite innocent of paper.

As not a ray of light relieved the darkness I put my hand into my ticket pocket, where I was accustomed to carry matches, and finding that my captors had not deprived me of them, lit one and looked about me. It was a dismal scene that little gleam illumined. The room in which I was confined was a small one, being only about ten feet long by eight wide, while, if I had been able to stand upright, I might have raised my hand to within two or three inches of the ceiling. In the furthest left-hand corner was a door, while in the wall on the right, but hopelessly beyond my reach, was a low window almost completely boarded up. I had no opportunity of seeing more, for by the time I had realized these facts the match had burnt down to my fingers. I blew it out and hastened to light another.

Just as I did so a low moan reached my ear. It came from the further end of the room. Again I held the match aloft; this time to discover a huddled-up figure in the corner opposite the door. One glance at it told me that it was none other than my young friend the Marquis of Beckenham. He was evidently still unconscious, for though I called him twice by name, he did not answer, but continued in the same position, moaning softly as before. I had only time for a hurried glance at him before my last match burned down to my fingers, and had to be extinguished. With the departure of the light a return of faintness seized me, and I fell back into my corner, if not quite insensible, certainly unconscious of the immediate awkwardness of our position.

It was daylight when my power of thinking returned to me, and long shafts of sunshine were percolating into us through the chinks in the boards upon the window. To my dismay the room looked even smaller and dingier than when I had examined it by the light of my match some hours before. The young Marquis lay unconscious in his corner just as I had last seen him, but with the widening light I discovered that his curious posture was due more to extraneous circumstances than to his own weakness, for I could see that he was fastened to the wall by a similar collar to my own.

I took out my watch, which had not been taken from me as I might have expected, and examined the dial. It wanted five minutes of six o'clock. So putting it back into my pocket, I set myself for the second time to try and discover where we were. By reason of my position and the chain that bound me, this could only be done by listening, so I shut my eyes and put all my being into my ears. For some moments no sound rewarded my attention. Then a cock in a neighbouring yard on my right crowed lustily, a dog on my left barked, and a moment later I heard the faint sound of someone coming along the street. The pedestrian, whoever he might be, was approaching from the right hand, and, what was still more important, my trained ear informed me that he was lame of one leg, and walked with crutches. Closer and closer he came. But to my surprise he did not pass the window; indeed, I noticed that when he came level with it the sound was completely lost to me. This told me two things: one, that the window, which, as I have already said, was boarded up, did not look into the main thoroughfare; the other, that the street itself ran along on the far side of the very wall to which my chain was attached.

As I arrived at the knowledge of this fact, Beckenham opened his eyes; he sat up as well as his chain would permit, and gazed about him in a dazed fashion. Then his right hand went up to the iron collar enclosing his neck, and when he had realized what it meant he appeared even more mystified than before. He seemed to doze again for a minute or so, then his eyes opened, and as they did so they fell upon me, and his perplexity found relief in words.

'Mr Hatteras,' he said, in a voice like that of a man talking in his sleep, 'where are we and what on earth does this chain mean?'

'You ask me something that I want to know myself,' I answered. 'I cannot tell you where we are, except that we are in Port Said. But if you want to know what I think it means, well, I think it means treachery. How do you feel now?'

'Very sick indeed, and my head aches horribly. But I can't understand it at all. What do you mean by saying that it is treachery?'

This was the one question of all others I had been dreading, for I could not help feeling that when all was said and done I was bitterly to blame. However, unpleasant or not, the explanation had to be got through, and without delay.

'Lord Beckenham,' I began, sitting upright and clasping my hands round my knees, 'this is a pretty bad business for me. I haven't the reputation of being a coward, but I'll own I feel pretty rocky and mean when I see you sitting there on the floor with that iron collar round your neck and that chain holding you to the wall, and know that it's, in a measure, all my stupid, blundering folly that has brought it about.'

'Oh, don't say that, Mr Hatteras!' was the young man's generous reply. 'For whatever or whoever may be to blame for it, I'm sure you're not.'

'That's because you don't know everything, my lord. Wait till you have heard what I have to tell you before you give me such complete absolution.'

'I'm not going to blame you whatever you may tell me; but please go on!'

There and then I set to work and told him all that had happened to me since my arrival in London; informed him of my meeting with Nikola, of Wetherell's hasty departure for Australia, of my distrust for Baxter, described the telegram incident and Baxter's curious behaviour afterwards, narrated my subsequent meeting with the two men in the Green Sailor Hotel, described my journey to Plymouth, and finished with the catastrophe that had happened to me there.

'Now you see,' I said in conclusion, 'why I regard myself as being so much to blame.'

'Excuse me,' he answered, 'but I cannot say that I see it in the same light at all.'

'I'm afraid I must be more explicit then. In the first place you must understand that, without a shadow of a doubt, Baxter was chosen for your tutor by Nikola, whose agent he undoubtedly is, for a specific purpose. Now what do you think that purpose was? You don't know? To induce your father to let you travel, to be sure. You ask why they should want you to travel? We'll come to that directly. Their plan is succeeding admirably, when I come upon the scene and, like the great blundering idiot I am, must needs set to work unconsciously to assist them in their nefarious designs. Your father eventually consents, and it is arranged that you shall set off for Australia at once. Then it is discovered that I am going to leave in the same boat. This does not suit Nikola's plans at all, so he determines to prevent my sailing with you. By a happy chance he is unsuccessful, and I follow and join the boat in Naples. Good gracious! I see something else now.'

'What is that?'

'Simply this. I could not help thinking at the time that your bout of seasickness between Naples and this infernal place was extraordinary. Well, if I'm not very much mistaken, *you were physicked, and it was Baxter's doing*.'

'But why?'

'Ah! That's yet to be discovered. But you may bet your bottom dollar it was some part of their devilish conspiracy. I'm as certain of that as that we are here now. Now here's another point. Do you remember my running out of the Casino last night? Well, that was because I saw Nikola standing in the roadway watching us.'

'Are you certain? How could he have got here? And what could his reasons be for watching us?'

'Why, can't you see? To find out how his plot is succeeding, to be sure.'

'And that brings us back to our original question—what is that plot?'

'That's rather more difficult to answer! But if you ask my candid opinion I should say nothing more nor less than to make you prisoner and blackmail your father for a ransom.'

For some few minutes neither of us spoke. The outlook seemed too hopeless for words, and the Marquis was still too weak to keep up an animated conversation for any length of time. He sat leaning his head on his hand. But presently he looked up again.

'My poor father!' he said. 'What a state he will be in!'

'And what worries me more,' I answered, 'is how he will regret ever having listened to my advice. What a dolt I was not to have told him of my suspicions.'

'You must not blame yourself for that. I am sure my father would hold you as innocent as I do. Now let us consider our position. In the first place, where are we, do you think? In the second, is there any possible chance of escape?'

'To the first my answer is, "don't know"; to the second, "can't say". I have discovered one thing, however, and that is that the street does not lie outside that window, but runs along on the other side of this wall behind me. The window, I suspect, looks out on to some sort of a courtyard. But unfortunately that information is not much use to us, as we can neither of us move away from where we are placed.'

'Is there no other way?'

'Not one, as far as I can tell. Can you see anything on your side?'

'Nothing at all, unless we could get at the door. But what's that sticking out of the wall near your feet?'

To get a better view of it I stooped as much as I was able.

'It looks like a pipe.'

The end of a pipe it certainly was, and sticking out into the room, but where it led to, and why it had been cut off in this peculiar fashion, were two questions I could no more answer than I could fly.

'Does it run out into the street, do you think?' was Beckenham's immediate query. 'If so, you might manage to call through it to some passer-by, and ask him to obtain assistance for us!'

'A splendid notion if I could get my mouth anywhere within a foot of it, but as this chain will not permit me to do that, it might as well be a hundred miles off. It's as much as I can do to touch it with my fingers.'

'Do you think if you had a stick you could push a piece of paper through? We might write a message on it.'

'Possibly, but there's another drawback to that. I haven't the necessary piece of stick.'

'Here is a stiff piece of straw; try that.'

He harpooned a piece of straw, about eight inches long, across the room towards me, and, when I had received it, I thrust it carefully into the pipe. A disappointment, however, was in store for us.

'It's no use,' I reported sorrowfully, as I threw the straw away. 'It has an elbow half-way down, and that would prevent any message from being pushed through.'

'Then we must try to discover some other plan. Don't lose heart!'

'Hush! I hear somebody coming.'

True enough a heavy footfall was approaching down the passage. It stopped at the door of the room in which we were confined, and a key was inserted in the lock. Next moment the door swung open and a tall man entered the room. A ray of sunlight, penetrating between the boards that covered the window, fell upon him, and showed us that his hair was white and that his face was deeply pitted with smallpox marks. Now, where had I met or heard of a man with those two peculiarities before? Ah! I remembered!

He stood for a moment in the doorway looking about him, and then strolled into the centre of the room.

'Good-morning, gentlemen,' he said, with an airy condescension that stung like an insult; 'I trust you have no fault to find with the lodging our poor hospitality is able to afford you.'

'Mr Prendergast,' I answered, determined to try him with the name of the man mentioned by my sweetheart in her letter. 'What does this mean? Why have we been made prisoners like this? I demand to be released at once. You will have to answer to our consul for this detention.'

For a brief space he appeared to be dumbfounded by my knowledge of his name. But he soon recovered himself and leaned his back against the wall, looking us both carefully over before he answered.

'I shall be only too pleased,' he said sneeringly, 'but if you'll allow me to say so, I don't think we need trouble about explanations yet awhile.'

'Pray, what do you mean by that?'

'Exactly what I say; as you are likely to be our guests for some considerable time to come, there will be no need for explanation.'

'You mean to keep us prisoners, then, do you? Very well, Mr Prendergast, be assured of this, when I *do* get loose I'll make you feel the weight of my arm.'

'I think it's very probable there will be a fight if ever we do meet,' he answered, coolly taking a cigarette from his pocket and lighting it. 'And it's my impression you'd be a man worth fighting, Mr Hatteras.'

'If you think my father will let me remain here very long you're much mistaken,' said Beckenham. 'And as for the ransom you expect him to pay, I don't somehow fancy you'll get a halfpenny.'

At the mention of the word 'ransom' I noticed that a new and queer expression came into our captor's face. He did not reply, however, except to utter his usual irritating laugh. Having done so he went to the door and called something in Arabic. In answer a gigantic negro made his appearance, bearing in his hands a tray on which were set two basins of food and two large mugs of water. These were placed before us, and Prendergast bade us, if we were hungry, fall to.

'You must not imagine that we wish to starve you,' he said. 'Food will be served to you twice a day. And if you want it, you can even be supplied with spirits and tobacco. Now, before I go, one word of advice. Don't indulge in any idea of escape. Communication with the outside world is absolutely impossible, and you will find that those collars and chains will stand a good strain before they will give way. If you behave yourselves you will be well looked after; but if you attempt any larks you will be confined in different rooms, and there will be a radical change in our behaviour towards you.'

So saying he left the room, taking the precaution to lock the door carefully behind him.

When we were once more alone, a long silence fell upon us. It would be idle for me to say that the generous behaviour of the young Marquis with regard to my share in this wretched business had set my mind at rest. But if it had not done that it had at least served to intensify another resolution. Come what might, I told myself, I would find a way of escape, and he should be returned to his father safe and sound, if it cost me my life to do it. But how *were* we to escape? We could not move from our places on account of the chains that secured us to the walls, and, though I put all my whole strength into it, I found

I could not dislodge the staple a hundredth part of an inch from its holding-place.

The morning wore slowly on, midday came and went, the afternoon dragged its dismal length, and still there was no change in our position. Towards sundown the same gigantic negro entered the room again, bringing us our evening meal. When he left we were locked up for the night, with only the contemplation of our woes, and the companionship of the multitudes of mice that scampered about the floor, to enliven us.

The events of the next seven days are hardly worth chronicling, unless it is to state that every morning at daylight the same cock crew and the same dog barked, while at six o'clock the same cripple invariably made his way down the street behind me. At eight o'clock, almost to the minute, breakfast was served to us, and, just as punctually, the evening meal made its appearance as the sun was declining behind the opposite housetop. Not again did we see any sign of Mr Prendergast, and though times out of number I tugged at my chain I was never a whit nearer loosening it than I had been on the first occasion. One after another plans of escape were proposed, discussed, and invariably rejected as impracticable. So another week passed and another, until we had been imprisoned in that loathsome place not less than twenty days. By the end of that time, as may be supposed, we were as desperate as men could well be. I must, however, admit here that anything like the patience and pluck of my companion under such trying circumstances I had never in my life met with before. Not once did he reproach me in the least degree for my share in the wretched business, but took everything just as it came, without unnecessary comment and certainly without complaint.

One fact had repeatedly struck me as significant, and that was the circumstance that every morning between six and half-past, as already narrated, the same cripple went down the street; and in connection with this, within the last few days of the time, a curious coincidence had revealed itself to me. From the tapping of his crutches on the stones I discovered that while one was shod with iron, the other was not. Now where and when had I noticed that peculiarity in a cripple before? That I had observed it somewhere I felt certain. For nearly half the day I turned this over and over in my mind, and then, in the middle of our evening meal, enlightenment came to me.

I remembered the man whose piteous tale had so much affected Beckenham on the day of our arrival, and the sound his crutches made upon the pavement as he left us. If my surmise proved correct, and we could only manage to communicate with him, here was a golden opportunity. But how were we to do this? We discussed it, and discussed it, times out of number, but in vain. That he must be stopped on his way down the street need not to be argued at all. In what way, however, could this be done? The window was out of the question, the door was not to be thought of; in that case the only communicating place would be the small pipe by my side. But as I have already pointed out, by reason of the elbow it would be clearly impossible to force a message through it. All day we devoted ourselves to attempts to solve what seemed a hopeless difficulty. Then like a flash another brilliant inspiration burst upon me.

'By Jove, I have it!' I said, taking care to whisper lest any one might be listening at the door. 'We must manage by hook or crook to catch a mouse *and let him carry our appeal for help to the outside world.*'

'A magnificent idea! If we can catch one I do believe you've saved us!'

But to catch a mouse was easier said than done. Though the room was alive with them they were so nimble and so cunning, that, try how we would, we could not lay hold of one. But at length my efforts were rewarded, and after a little struggle I held my precious captive in my hand. By this time another idea had come to me. If we wanted to bring Nikola and his gang to justice, and to discover their reason for hatching this plot against us, it would not do to ask the public at large for help—and I must own, in spite of our long imprisonment, I was weak enough to feel a curiosity as to their motive. No! It must be to the beggar who passed the house every morning that we must appeal.

'This letter concerns you more than me,' I said to my fellow prisoner. 'Have you a lead pencil in your pocket?'

He had, and immediately threw it across to me. Then, taking a small piece of paper from my pocket, I set myself to compose the following in French and English, assisted by my companion:

If this should meet the eye of the individual to whom a young Englishman gave half a sovereign in charity three weeks ago, he is implored to assist one who assisted him, and who has been imprisoned ever since that day in the room with the blank wall facing the street and the boarded-up window on the right hand side. To do this he must obtain a small file and discover a way to convey it into the room by means of the small pipe leading through the blank wall into the street; perhaps if this could be dislodged it might be pushed in through the aperture thus made. On receipt of the file an English five-pound note will be conveyed to him in the same way as this letter, and another if secrecy is observed and those imprisoned in the house escape.

This important epistle had hardly been concocted before the door was unlocked and our dusky servitor entered with the evening meal. He had long since abandoned his first habit of bringing us our food in separate receptacles, but conveyed it to us now in the saucepan in which it was cooked, dividing it thence into our basins. These latter, it may be interesting to state, had not been washed since our arrival.

All the time that our jailer was in the room I held my trembling prisoner in my hand, clinging to him as to the one thing which connected us with liberty. But the door had no sooner closed upon him than I had tilted out my food upon the floor and converted my basin into a trap.

It may be guessed how long that night seemed to us, and with what trembling eagerness we awaited the first signs of breaking day. Directly it was light I took off and unravelled one of my socks. The thread thus obtained I doubled, and having done this, secured one end of it to the note, which I had rolled into a small compass, attaching the other to my captive mouse's hind leg. Then we set ourselves to wait for six o'clock. The hour came; and minute after minute went by before we heard in the distance the tapping of the crutches on the stones. Little by little the sound grew louder, and then fainter, and when I judged he was nearly at my back I stooped and thrust our curious messenger into the pipe. Then we sat down to await the result.

As the mouse, only too glad to escape, ran into the aperture, the thread, on which our very lives depended, swiftly followed, dragging its message after it. Minutes went by; half an hour; an hour; and then the remainder of the day; and still nothing came to tell us that our appeal had been successful.

That night I caught another mouse, wrote the letter again, and at six o'clock next morning once more dispatched it on its journey. Another day went by without reply. That night we caught another, and at six o'clock next morning sent it off; a third, and even a fourth, followed, but still without success. By this time the mice were almost impossible to catch, but our wits were sharpened by despair, and we managed to hit upon a method that eventually secured for us a plentiful supply. For the sixth time the letter was written and dispatched at the moment the footsteps were coming down the street. Once more the tiny animal crawled into the pipe, and once more the message disappeared upon its journey.

Another day was spent in anxious waiting, but this time we were not destined to be disappointed. About eight o'clock that night, just as we were giving up hope, I detected a faint noise near my feet; it was for all the world as if some one were forcing a stick through a hole in a brick wall. I informed Beckenham of the fact in a whisper, and then put my head down to listen. Yes, there was the sound again. Oh, if only I had a match! But it was no use wishing for what was impossible, so I put my hand down to the pipe. *It was moving!* It turned in my hand, moved to and fro for a brief space and then disappeared from my grasp entirely; next moment it had left the room. A few seconds later something cold was thrust into my hand, *and from its rough edge I knew it to be a file.* I drew it out as if it were made of gold and thrust it into my pocket. A piece of string was attached to it, and the reason of this I was at first at some loss to account for. But a moment's reflection told me that it was to assist in the fulfilment of our share of the bargain. So, taking a five-pound note from the secret pocket in which I carried my paper money, I tied the string to it, and it was instantly withdrawn.

A minute could not have elapsed before I was at work upon the staple of my collar, and in less than half an hour it was filed through and the iron was off my neck.

If I tried for a year I could not make you understand what a relief it was to me to stand upright. I stretched myself again and again, and then crossed the room on tiptoe in the dark to where the Marquis lay.

'You are free!' he whispered, clutching and shaking my hand. 'Oh, thank God!'

'Hush! Put down your head and let me get to work upon your collar before you say anything more.'

As I was able this time to get at my work standing up, it was not very long before Beckenham was as free as I was. He rose to his feet with a great sigh of relief, and we shook hands warmly in the dark.

'Now,' I said, leading him towards the door, 'we will make our escape, and I pity the man who attempts to stop us.'

CHAPTER IX

Dr Nikola permits us a Free Passage

The old saying, 'Don't count your chickens before they're hatched,' is as good a warning as any I know. Certainly it proved so in our case. For if we had not been so completely occupied filing through the staples of our collars we should not have omitted to take into consideration the fact that, even when we should have removed the chains that bound us, we would still be prisoners in the room. I'm very much afraid, however, even had we remembered this point, we should only have considered it of minor importance and one to be easily overcome. As it was, the unwelcome fact remained that the door *was* locked, and, what was worse, that the lock itself had, for security's sake, been placed on the outside, so that there was no chance of our being able to pick it, even had our accomplishments lain in that direction.

'Try the window,' whispered Beckenham, in answer to the heavy sigh which followed my last discovery.

Accordingly we crossed the room, and I put my hands upon one of the boards and pulled. But I might as well have tried to tow a troopship with a piece of cotton, for all the satisfactory result I got; the planks were trebly screwed to the window frame, and each in turn defied me. When I was tired Beckenham put his strength to it, but even our united efforts were of no avail, and, panting and exhausted, we were at length obliged to give it up as hopeless.

'This is a pretty fix we've got ourselves into,' I said as soon as I had recovered sufficient breath to speak. 'We can't remain here, and yet how on earth are we to escape?'

'I can't say, unless we manage to burst that door open and fight our way out. I wonder if that could be done.'

'First, let's look at the door.'

We crossed the room again, and I examined the door carefully with my fingers. It was not a very strong one; but I was sufficient of a carpenter to know that it would withstand a good deal of pressure before it would give way.

'I've a good mind to try it,' I said; 'but in that case, remember, it will probably mean a hand-to-hand fight on the other side, and, unarmed and weak as we are, we shall be pretty sure to get the worst of it.'

'Never mind that,' my intrepid companion replied, with a confidence in his voice that I was very far from feeling. 'In for a penny, in for a pound; even if we're killed it couldn't be worse than being buried alive in here.'

'That's so, and if fighting's your idea, I'm your man,' I answered. 'Let me first take my bearings, and then I'll see what I can do against it. You get out of the way, but be sure to stand by to rush the passage directly the door goes.'

Again I felt the door and wall in order that I might be sure where it lay, and having done so crossed the room. My heart was beating like a Nasmyth hammer, and it was nearly a minute before I could pull myself together sufficiently for my rush. Then summoning every muscle in my body to my assistance, I dashed across and at it with all the strength my frame was capable of. Considering the darkness of the room, my steering was not so bad, for my shoulder caught the door just above its centre; there was a great crash—a noise of breaking timbers—and amid a shower of splinters and general debris I fell headlong through into the passage. By the time it would have taken me to count five, Beckenham was beside me helping me to rise.

'Now stand by for big trouble!' I said, rubbing my shoulder, and every moment expecting to see a door open and a crowd of Prendergast's ruffians come rushing out. 'We shall have them on us in a minute.'

But to our intense astonishment it was all dead silence. Not a sound of any single kind, save our excited breathing, greeted our ears. We might have broken into an empty house for all we knew the difference.

For nearly five minutes we stood, side by side, waiting for the battle which did not come.

'What on earth does it mean?' I asked my companion. 'That crash of mine was loud enough to wake the dead. Can they have deserted the place, think you, and left us to starve?'

'I can't make it out any more than you can,' he answered. 'But don't you think we'd better take advantage of their not coming to find a way out?'

'Of course. One of us had better creep down the passage and discover how the land lies. As I'm the stronger, I'll go. You wait here.'

Dr Nikola permits us a Free Passage

I crept along the passage, treading cautiously as a cat, for I knew that both our lives depended on it. Though it could not have been more than sixty feet, it seemed of interminable length, and was as black as night. Not a glimmer of light, however faint, met my eyes.

On and on I stole, expecting every moment to be pounced upon and seized; but no such fate awaited me. If, however, our jailers did not appear, another danger was in store for me.

In the middle of my walk my feet suddenly went from under me, and I found myself falling I knew not where. In reality it was only a drop of about three feet down a short flight of steps. Such a noise as my fall made, however, was surely never heard, but still no sound came. Then Beckenham fumbled his way cautiously down the steps to my side, and whispered an enquiry as to what had happened. I told him in as few words as possible, and then struggled to my feet again.

Just as I did so my eyes detected a faint glimmer of light low down on the floor ahead of us. From its position it evidently emanated from the doorway of a room.

'Oh! if only we had a match,' I whispered.

'It's no good wishing,' said Beckenham. 'What do you advise?'

'It's difficult to say,' I answered; 'but I should think we'd better listen at that door and try to discover if there is any one inside. If there is, and he is alone, we must steal in upon him, let him see that we are desperate, and, willy-nilly, force him to show us a way out. It's ten chances to one, if we go on prowling about here, we shall stumble upon the whole nest of them—then we'll be caught like rats in a trap. What do you think?'

'I agree with you. Go on.'

Without further ado we crept towards the light, which, as I expected, came from under a door, and listened. Someone was plainly moving about inside; but though we waited for what seemed a quarter of an hour, but must in reality have been less than a minute and a half, we could hear no voices.

'Whoever he is, he's alone—that's certain,' whispered my companion. 'Open the door softly, and we'll creep in upon him.'

In answer, and little by little, a cold shiver running down my back lest it should creak and so give warning to the person within, I turned the handle, pushed open the door, and we looked inside. Then—but, my gracious! if I live to be a thousand I shall never forget even the smallest particular connected with the sight that met my eyes.

The room itself was a long and low one: its measurements possibly sixty feet by fifteen. The roof—for there was no ceiling—was of wood, crossed by heavy rafters, and much begrimed with dirt and smoke. The floor was of some highly polished wood closely resembling oak and was completely bare. But the shape and construction of the room itself were as nothing compared with the strangeness of its furniture and occupants. Words would fail me if I tried to give you a true and accurate description of it. I only know that, strong man as I was, and used to the horrors of life and death, what I saw before me then made my blood run cold and my flesh creep as it had never done before.

To begin with, round the walls were arranged, at regular intervals, more than a dozen enormous bottles, each of which contained what looked, to me, only too much like human specimens pickled in some light-coloured fluid resembling spirits of wine. Between these gigantic but more than horrible receptacles were numberless smaller ones holding other and even more dreadful remains; while on pedestals and stands, bolt upright and reclining, were skeletons of men, monkeys, and quite a hundred sorts of animals. The intervening spaces were filled with skulls, bones, and the apparatus for every kind of murder known to the fertile brain of man. There were European rifles, revolvers, bayonets, and swords; Italian stilettos, Turkish scimitars, Greek knives, Central African spears and poisoned arrows, Zulu knobkerries, Afghan yataghans, Malay krises, Sumatra blowpipes, Chinese dirks, New Guinea head-catching implements, Australian spears and boomerangs, Polynesian stone hatchets, and numerous other weapons the names of which I cannot now remember. Mixed up with them were implements for every sort of wizardry known to the superstitious; from old-fashioned English love charms to African Obi sticks, from spiritualistic planchettes to the most horrible of Fijian death potions.

In the centre of the wall, opposite to where we stood, was a large fireplace of the fashion usually met with in old English manor-houses, and on either side of it a figure that nearly turned me sick with horror. That on the right hand was apparently a native of Northern India, if one might judge by his dress and complexion. He sat on the floor in a constrained attitude, accounted for by the fact that his head, which was at least three times too big for his body, was so heavy as to require an iron tripod with a ring or collar in the top of it to keep it from overbalancing him and bringing him to the floor. To add to the horror of this awful head, it was quite bald; the skin was drawn tensely over the bones, and upon this great veins stood out as large as macaroni stems.

On the other side of the hearth was a creature half-ape and half-man—the like of which I remember once to have seen in a museum of monstrosities in Sydney, where, if my memory serves me, he was described upon the catalogue as a Burmese monkey-boy. He was chained to the wall in somewhat the same fashion as we had been, and was chattering and scratching for all the world like a monkey in a Zoo.

But, horrible as these things were, the greatest surprise of all was yet to come. For, standing at the heavy oaken table in the centre of the room, was a man I should have known anywhere if I had been permitted half a glance at him. *It was Dr Nikola.*

When we entered he was busily occupied with a scalpel, dissecting an animal strangely resembling a monkey. On the table, and watching the work upon which his master was engaged, sat his constant companion, the same fiendish black cat I have mentioned elsewhere. While at the end nearest us, standing on tiptoe, the better to see what was going on, was an albino dwarf, scarcely more than two feet eight inches high.

Now, though it has necessarily taken me some time to describe the scene which greeted our eyes, it must not be supposed that anything like the same length of time had really elapsed since our entry. Three seconds at the very most would have sufficed to cover the whole period.

So stealthily, however, had our approach been made, and so carefully had I opened the door, that we were well into the room before our appearance was discovered, and also before I had realized into whose presence we had stumbled. Then my foot touched a board that creaked, and Dr Nikola looked up from the work upon which he was engaged.

His pale, thin face did not show the slightest sign of surprise as he said, in his usual placid tone,

'So you have managed to escape from your room, gentlemen. Well, and pray what do you want with me?'

For a moment I was so much overcome with surprise that my tongue refused to perform its office. Then I said, advancing towards him as I spoke, closely followed by the Marquis,

'So, Dr Nikola, we have met at last!'

'At last, Mr Hatteras, as you say,' this singular being replied, still without showing a sign of either interest or embarrassment. 'All things considered, I suppose you would deem me ironical if I ventured to say that I am pleased to see you about again. However, don't let me keep you standing, won't you sit down? My lord, let me offer you a chair.'

All this time we were edging up alongside the table, and I was making ready for a rush at him. But he was not to be taken off his guard. His extraordinary eyes had been watching me intently, taking in my every movement; and a curious effect their steady gaze had upon me.

'Dr Nikola,' I said, pulling myself together, 'the game is up. You beat me last time; but now you must own I come out on top. Don't utter a word or call for assistance—if you do you're a dead man. Now drop that knife you hold in your hand, and show us the way out!'

The Marquis was on his right, I was on his left, and we were close upon him as I spoke. Still he showed no sign of fear, though he must have known the danger of his position. But his eyes glowed in his head like living coals.

You will ask why we did not rush at him? Well, if I am obliged to own it, I must—the truth was, such was the power that emanated from this extraordinary man, that though we both knew the crucial moment of our enterprise had arrived, while his eyes were fixed upon us, neither of us could stir an inch. When he spoke his voice seemed to cut like a knife.

'So you think my game is up, Mr Hatteras, do you? I'm afraid once more I must differ from you. Look behind you.'

I did so, and that glance showed me how cleverly we'd been trapped. Leaning against the door, watching us with cruel, yet smiling eyes, was our old enemy Prendergast, revolver in hand. Just behind me were two powerful Soudanese, while near the Marquis was a man looking like a Greek—and a very stalwart Greek at that. Observing our discomfiture, Nikola seated himself in a big chair near the fireplace and folded his hands in the curious fashion I have before

Dr Nikola permits us a Free Passage

described; as he did so his black cat sprang to his shoulder and sat there watching us all. Dr Nikola was the first to speak.

'Mr Hatteras,' he said, with devilish clearness and deliberation, 'you should really know me better by this time than to think you could outwit me so easily. Is my reputation after all so small? And, while I think of it, pray let me have the pleasure of returning to you your five pound note and your letters. Your mice were perfect messengers, were they not?' As he spoke he handed me the self-same Bank of England note I had dispatched through the pipe that very evening in payment for the file; then he shook from a box he had taken from the chimney-piece all the communications I had written imploring assistance from the outside world. To properly estimate my chagrin and astonishment would be very difficult. I could only sit and stare, first at the money and then at the letters, in blankest amazement. So we had not been rescued by the cripple after all. Was it possible that while we had been so busy arranging our escape we had in reality been all the time under the closest surveillance? If that were so, then this knowledge of our doings would account for the silence with which my attack upon the door had been received. Now we were in an even worse position than before. I looked at Beckenham, but his head was down and his right hand was picking idly at the table edge. He was evidently waiting for what was coming next. In sheer despair I turned to Nikola.

'Since you have outwitted us again, Dr Nikola, do not play with us—tell us straight out what our fate is to be.'

'If it means going back to that room again,' said Beckenham, in a voice I hardly recognized, 'I would far rather die and be done with it.'

'Do not fear, my lord, you shall not die,' Nikola said, turning to him with a bow. 'Believe me, you will live to enjoy many happier hours than those you have been compelled to spend under my roof!'

'What do you mean?'

The doctor did not answer for nearly a moment; then he took what looked to me suspiciously like a cablegram form from his pocket and carefully examined it. Having done so, he said quietly,

'Gentlemen, you ask what I mean? Well, I mean this—if you wish to leave this house this very minute, you are free to do so on one condition!'

'And that condition is?'

'That you allow yourselves to be blindfolded in this room and conducted by my servants to the harbour side. I must furthermore ask your words of honour that you will not seek to remove your bandages until you are given permission to do so. Do you agree to this?'

Needless to say we both signified our assent.

This free permission to leave the house was a second surprise, and one for which we were totally unprepared.

'Then let it be so. Believe me, my lord Marquis, and you, Mr Hatteras, it is with the utmost pleasure I restore your liberty to you again!'

He made a sign to Prendergast, who instantly stepped forward. But I had something to say before we were removed.

'One word first, Dr Nikola. You have——'

'Mr Hatteras, if you will be guided by me, you will keep a silent tongue in your head. Let well alone. Take warning by the proverb, and beware how you disturb a sleeping dog. Why I have acted as I have done towards you, you may some day learn; in the meantime rest assured it was from no idle motive. Now take me at my word, and go while you have the chance. I may change my mind in a moment, and then——'

He stopped and did not say any more. At a sign, Prendergast clapped a thick bandage over my eyes, while another man did the same for Beckenham; a man on either side of me took my arms, and next moment we had passed out of the room, and before I could have counted fifty were in the cool air of the open street.

How long we were walking, after leaving the house, I could not say, but at last our escort called a halt. Prendergast was evidently in command, for he said,

'Gentlemen, before we leave you, you will renew your words of honour not to remove your bandages for five full minutes?'

We complied with his request, and instantly our arms were released; a moment later we heard our captors leaving us. The minutes went slowly by. Presently Beckenham said,

'How long do you think we've been standing here?'

'Nearly the stipulated time, I should fancy,' I answered. 'However, we'd better give them a little longer, to avoid any chance of mistake.'

Again a silence fell on us. Then I tore off my bandage, to find Beckenham doing the same.

'They're gone, and we're free again,' he cried. 'Hurrah!'

We shook hands warmly on our escape, and having done so looked about us. A ship's bell out in the stream chimed half an hour after midnight, and a precious dark night it was. A number of vessels were to be seen, and from the noise that came from them it was evident they were busy coaling.

'What's to be done now?' asked Beckenham.

'Find an hotel, I think,' I answered; 'get a good night's rest, and first thing in the morning hunt up our consul and the steamship authorities.'

'Come along, then. Let's look for a place. I noticed one that should suit us close to where we came ashore that day.'

Five minutes' walking brought us to the house we sought. The proprietor was not very fastidious, and whatever he may have thought of our appearances he took us in without demur. A bath and a good meal followed, and then after a thorough overhauling of all the details connected with our imprisonment we turned into bed, resolved to thrash it out upon the morrow.

Next morning, true to our arrangement, as soon as breakfast was over, I set off for the steamship company's office, leaving the Marquis behind me at the hotel for reasons which had begun to commend themselves to me, and which will be quite apparent to you.

I found the *Saratoga*'s agent hard at work in his private office. He was a tall, thin man, slightly bald, wearing a pair of heavy gold pince-nez, and very slow and deliberate in his speech.

'I beg your pardon,' he began, when I had taken possession of his proffered chair, 'but did I understand my clerk to say that your name was Hatteras?'

'That is my name,' I answered. 'I was a passenger in your boat the *Saratoga* for Australia three weeks ago, but had the misfortune to be left behind when she sailed.'

'Ah! I remember the circumstances thoroughly,' he said. 'The young Marquis of Beckenham went ashore with you, I think, and came within an ace of being also left behind.'

'Within an ace!' I cried; 'but he *was* left behind.'

'No, no! there you are mistaken,' was the astounding reply; 'he *would* have been left behind had not his tutor and I gone ashore at the last moment to look for him and found him wandering about on the outskirts of Arab Town. I don't remember ever to have seen a man more angry than the tutor was, and no wonder, for they only just got out to the boat again as the gangway was being hauled aboard.'

'Then you mean to tell me that the Marquis went on to Australia after all!' I cried. 'And pray how did this interesting young gentleman explain the fact of his losing sight of me?'

'He lost you in a crowd, he said,' the agent continued. 'It was a most extraordinary business altogether.'

It certainly was, and even more extraordinary than he imagined. I could hardly believe my ears. The world seemed to be turned upside down. I was so bewildered that I stumbled out a few lame enquiries about the next boat sailing for Australia, and what would be done with my baggage on its arrival at the other end, and then made my way as best I could out of the office.

Hastening back to the hotel, I told my story from beginning to end to my astonished companion, who sat on his bed listening open-mouthed. When I had finished he said feebly,

'But what does it all mean? Tell me that! What does it mean?'

'It means,' I answered, 'that our notion about Nikola's abducting us in order to blackmail your father was altogether wrong, and, if you ask me, I should say not half picturesque enough. No, no! this mystery is a bigger one by a hundred times than even we expected, and there are more men in it than those we have yet seen. It remains with you to say whether you will assist in the attempt to unravel it or not.'

'What do you mean by saying it remains with me? Do I understand that you intend following it up?'

'Of course I do. Nikola and Baxter between them have completely done me—now I'm going to do my best to do them. By Jove!'

'What is it now?'

'I see it all as plain as a pikestaff. I understand exactly now why Baxter came for you, why he telegraphed that the train was laid, why I was drugged in Plymouth, why you were seasick between Naples and this place, and why we were both kidnapped so mysteriously!'

'Then explain, for mercy's sake!'

'I will. See here. In the first place, remember your father's peculiar education of yourself. If you consider that, you will see that you are the only young nobleman of high rank whose face is not well known to his brother peers. That being so, Nikola wants to procure you for some purpose of his own in Australia. Your father advertises for a tutor; he sends one of his agents—Baxter—to secure the position. Baxter, at Nikola's instruction, puts into your head a desire for travel. You pester your father for the necessary permission. Just as this is granted I come upon the scene. Baxter suspects me. He telegraphs to Nikola "The train is laid," which means that he has begun to sow the seeds of a desire for travel, when a third party steps in—in other words, I am the new danger that has arisen. He arranges your sailing, and all promises to go well. Then Dr Nikola finds out I intend going in the same boat. He tries to prevent me; and I—by Jove! I see another thing. Why did Baxter suggest that you should cross the Continent and join the boat at Naples? Why, simply because if you had started from Plymouth you would soon have got over your sickness, if you had ever been ill at all, and in that case the passengers would have become thoroughly familiar with your face by the time you reached Port Said. That would never have done, so he takes you to Naples, drugs you next morning—for you must remember you were ill after the coffee he gave you—and by that means keeps you ill and confined to your cabin throughout the entire passage to Port Said. Then he persuades you to go ashore with me. You do so, with what result you know. Presently he begins to bewail your non-return, invites the agent to help in the search. They set off, and eventually find you near the Arab quarter. You must remember that neither the agent, the captain, nor the passengers have seen you, save at night, so the substitute, who is certain to have been well chosen and schooled for the part he is to play, is not detected. Then the boat goes on her way, while we are left behind languishing in durance vile.'

'Do you really think those are the facts of the case?'

'Upon my word, I do!'

'Then what do you advise me to do? Remember, Baxter has letters to the different Governors from my father.'

'I know what I should do myself!'

'Go to the consul and get him to warn the authorities in Australia, I suppose?'

'No. That would do little or no good—remember, they've three weeks' start of us.'

'Then what shall we do? I'm in your hands entirely, and whatever you advise I promise you I'll do.'

'If I were you I should doff my title, take another name, and set sail with me for Australia. Once there, we'll put up in some quiet place and set ourselves to unmask these rascals and to defeat their little game, whatever it may be. Are you prepared for so much excitement as that?'

'Of course I am. Come what may, I'll go with you, and there's my hand on it.'

'Then we'll catch the next boat—not a mail-steamer—that sails for an Australian port, and once ashore there we'll set the ball a-rolling with a vengeance.'

'That scoundrel Baxter! I'm not vindictive as a rule, but I feel I should like to punish him.'

'Well, if they've not flown by the time we reach Australia, you'll probably be able to gratify your wish. It's Nikola, however, I want.'

Beckenham shuddered as I mentioned the Doctor's name. So to change the subject I said,

'I'm thinking of taking a little walk. Would you care to accompany me?'

'Where are you going?' he asked.

'I'm going to try and find the house where we were shut up,' I answered. 'I want to be able to locate it for future reference, if necessary.'

'Is it safe to go near it, do you think?'

'In broad daylight, yes! But, just to make sure, we'll buy a couple of revolvers on the way. And, what's more, if it becomes necessary we'll use them.'

'Come along, then.'

With that we left our hotel and set off in the direction of the Casino, stopping, however, on the way to make the purchases above referred to.

On arrival at the place we sought, we halted and looked about us. I pointed to a street on our right.

'That was the way we came from the mosque,' I said. Then, pointing to a narrow alley way almost opposite where we stood, I continued, 'And that was where I saw Nikola standing watching us. Now when we came out of this building we turned to our left hand, and, if I mistake not, went off in that direction. I think, if you've no objection, we'll go that way now.'

We accordingly set off at a good pace, and after a while arrived at the spot where the guide had caught us up. It looked a miserably dirty neighbourhood in the bright sunlight. Beckenham gazed about him thoughtfully, and finally said,

'Now we turn to our right, I think.'

'Quite so. Come along!'

We passed down one thoroughfare and up another, and at last reached the spot where I had commented on the sign-boards, and where we had been garrotted. Surely the house must be near at hand now? But though we hunted high and low, up one street and down another, not a single trace of any building, answering the description of the one we wanted, could we discover. At last, after nearly an hour's search, we were obliged to give it up, and return to our hotel, unsuccessful.

As we finished lunch a large steamer made her appearance in the harbour, and brought up opposite the town. When we questioned our landlord, who was an authority on the subject, he informed us that she was the s.s. *Pescadore*, of Hull, bound to Melbourne.

Hearing this we immediately chartered a boat, pulled off to her, and interviewed the captain. As good luck would have it, he had room for a couple of passengers. We therefore paid the passage money, went ashore again and provided ourselves with a few necessaries, rejoined her, and shortly before nightfall steamed into the Canal. Port Said was a thing of the past. Our eventful journey was resumed— what was the end of it all to be?

PART II

CHAPTER I

We reach Australia, and the Result

The *Pescadore*, if she was slow, was certainly sure, and so the thirty-sixth day after our departure from Port Said, as recorded in the previous chapter, she landed us safe and sound at Williamstown, which, as all the Australian world knows, is one of the principal railway termini, and within an hour's journey, of Melbourne. Throughout the voyage nothing occurred worth chronicling, if I except the curious behaviour of Lord Beckenham, who, for the first week or so, seemed sunk in a deep lethargy, from which neither chaff nor sympathy could rouse him. From morning till night he mooned aimlessly about the decks, had visibly to pull himself together to answer such questions as might be addressed to him, and never by any chance sustained a conversation beyond a few odd sentences. To such a pitch did this depression at last bring him that, the day after we left Aden, I felt it my duty to take him to task and to try to bully or coax him out of it. We were standing at the time under the bridge and a little forrard of the chart-room.

'Come,' I said, 'I want to know what's the matter with you. You've been giving us all the miserables lately, and from the look of your face at the present moment I'm inclined to believe it's going to continue. Out with it! Are you homesick, or has the monotony of this voyage been too much for you?'

He looked into my face rather anxiously, I thought, and then said:

'Mr Hatteras, I'm afraid you'll think me an awful idiot when I *do* tell you, but the truth is I've got Dr Nikola's face on my brain, and do what I will I cannot rid myself of it. Those great, searching eyes, as we saw them in that terrible room, have got on my nerves, and I can think of nothing else. They haunt me night and day!'

'Oh, that's all fancy!' I cried. 'Why on earth should you be frightened of him? Nikola, in spite of his demoniacal cleverness, is only a man, and even then you may consider that we've seen the last of him. So cheer up, take as much exercise as you possibly can, and believe me, you'll soon forget all about him.'

But it was no use arguing with him. Nikola had had an effect upon the youth that was little short of marvellous, and it was not until we had well turned the Leuwin, and were safely in Australian waters, that he in any way recovered his former spirits.

And here, lest you should give me credit for a bravery I did not possess, I must own that I was more than a little afraid of another meeting with Nikola, myself. I had had four opportunities afforded me of judging of his cleverness—once in the restaurant off Oxford Street, once in the Green Sailor public-house in the East India Dock Road, once in the West of England express, and lastly, in the house in Port Said. I had not the slightest desire, therefore, to come to close quarters with him again.

Arriving in Melbourne we caught the afternoon express for Sydney, reaching that city the following morning a little after breakfast. By the time we arrived at our destination we had held many consultations over our future, and the result was a decision to look for a quiet hotel on the outskirts of the city, and then to attempt to discover what the mystery, in which we had been so deeply involved, might mean. The merits of all the various suburbs were severally discussed, though I knew but little about them, and the Marquis less. Paramatta, Penrith, Woolahra, Balmain, and even many of the bays and harbours, received attention, until we decided on the last named as the most likely place to answer our purpose.

This settled, we crossed Darling harbour, and, after a little hunting about, discovered a small but comfortable hotel situated in a side street, called the 'General Officer'. Here we booked rooms, deposited our meagre baggage, and having installed ourselves, sat down and discussed the situation.

'So this is Sydney,' said Beckenham, stretching himself out comfortably upon the sofa by the window as he spoke. 'And now that we've got here, what's to be done first?'

'Have lunch,' I answered promptly.

'And then?' he continued.

'Hunt up a public library and take a glimpse of the *Morning Herald's* back numbers. They will tell us a good deal, though not all we want to know. Then we'll make a few enquiries. Tomorrow morning I shall ask you to excuse me for a couple of hours. But in the afternoon we ought to have acquired sufficient information to enable us to make a definite start on what we've got to do.'

'Then let's have lunch at once and be off. I'm all eagerness to get to work.'

We accordingly ordered lunch, and, when it was finished, set off in search of a public library. Having found it—and it was not a very difficult matter—we sought the reading room and made for a stand of *Sydney Morning Heralds* in the corner. Somehow I felt as certain of finding what I wanted there as any man could possibly be, and as it happened I was not disappointed. On the second page, beneath a heading in bold type, was a long report of a horse show, held the previous afternoon, at which it appeared a large vice-regal and fashionable party were present. The list included His Excellency the Governor and the Countess of Amberley, the Ladies Maud and Ermyntrude, their daughters, the Marquis of Beckenham, Captain Barrenden, an aide-de-camp, and Mr Baxter. In a voice that I hardly recognized as my own, so shaken was it with excitement, I called Beckenham to my side and pointed out to him his name. He stared, looked away, then stared again, hardly able to believe his eyes.

'What does it mean?' he whispered, just as he had done in Port Said. 'What does it mean?'

I led him out of the building before I answered, and then clapped him on the shoulder.

'It means, my boy,' I said, 'that there's been a hitch in their arrangements, and that we're not too late to circumvent them after all.'

'But where do you think they are staying—these two scoundrels?'

'At Government House, to be sure. Didn't you see that the report said, "The Earl and Countess of Amberley and a distinguished party from Government House, including the Marquis of Beckenham," etc.?'

'Then let us go to Government House at once and unmask them. That is our bounden duty to society.'

'Then all I can say is, if it is our duty to society, society will have to wait. No, no! We must find out first what their little game is. That once decided, the unmasking will fall in as a natural consequence. Don't you understand?'

'I am afraid I don't quite. However, I expect you're right.'

By this time we were back again at the ferry. It was not time for the boat to start, so while we waited we amused ourselves staring at the placards pasted about on the wharf hoardings. Then a large theatrical poster caught my eye and drew me towards it. It announced a grand vice-regal 'command' night at one of the principal theatres for that very evening, and further set forth the fact that the most noble the Marquis of Beckenham would be amongst the distinguished company present.

'Here we are,' I called to my companion, who was at a little distance. 'We'll certainly go to this. The Marquis of Beckenham shall honour it with his patronage and presence after all.'

Noting the name and address of the theatre, we went back to our hotel for dinner, and as soon as it was eaten returned to the city to seek the theatre.

When we entered it the building was crowded, and the arrival of the Government House party was momentarily expected. Presently there was a hush, then the orchestra and audience rose while 'God save the Queen' was played, and the Governor and a brilliant party entered the vice-regal box. You may be sure of all that vast concourse of people there were none who stared harder than Beckenham and myself. And it was certainly enough to make any man stare, for there, sitting on her ladyship's right hand, faultlessly dressed, was the exact image of the young man by my side. The likeness was so extraordinary that for a moment I could hardly believe that Beckenham had not left me to go up and take his seat there. And if I was struck by the resemblance, you may be sure that he was a dozen times more so. Indeed, his bewilderment was most comical, and must have struck those people round us, who were watching, as something altogether extraordinary. I looked again, and could just discern behind the front row the smug, self-satisfied face of the tutor Baxter. Then the play commenced, and we were compelled to turn and give it our attention.

Here I must stop to chronicle one circumstance that throughout the day had struck me as peculiar. When our vessel arrived at

We reach Australia, and the Result

Williamstown, it so happened that we had travelled up in the train to Melbourne with a tall, handsome, well-dressed man of about thirty years of age. Whether he, like ourselves, was a new arrival in the Colony, and only passing through Melbourne, I cannot say; at any rate he went on to Sydney in the mail train with us. Then we lost sight of him, only to find him standing near the public library when we had emerged from it that afternoon, and now here he was sitting in the stalls of the theatre not half a dozen chairs from us. Whether this continual companionship was designed or only accidental, I could not of course say, but I must own that I did not like the look of it. Could it be possible, I asked myself, that Nikola, learning our departure for Australia in the *Pescadore*, had cabled from Port Said to this man to watch us? It seemed hardly likely, and yet we had had sufficient experience of Nikola to teach us not to consider anything impossible for him to do.

The performance over, we left the theatre and set off for the ferry, only reaching it just as the boat was casting off. As it was I had to jump for it, and on reaching the deck should have fallen in a heap but for a helping hand that was stretched out to me. I looked up to tender my thanks, when to my surprise I discovered that my benefactor was none other than the man to whom I have just been referring. His surprise was even greater than mine, and muttering something about 'a close shave', he turned and walked quickly aft. My mind was now made up, and I accordingly reported my discovery to Beckenham, pointing out the man and warning him to watch for him when he was abroad without me. This he promised to do.

Next morning I donned my best attire (my luggage having safely arrived), and shortly before eleven o'clock bade Beckenham goodbye and betook myself to Potts Point to call upon the Wetherells.

It would be impossible for me to say with what varied emotions I trod that well-remembered street, crossed the garden, and approached the ponderous front door, which somehow had always seemed to me so typical of Mr Wetherell himself. The same butler who had opened the door to me on the previous occasion opened it now, and when I asked if Miss Wetherell were at home, he gravely answered, 'Yes, sir,' and invited me to enter. Though I had called there before, it must be remembered that this was the first time I had been inside the house, and I must confess the display of wealth in the hall amazed me.

I was shown into the drawing-room—a large double chamber beautifully furnished and possessing an elegantly painted ceiling—while the butler went in search of his mistress. A few moments later I heard a light footstep outside, a hand was placed upon the handle of the door, and before I could have counted ten, Phyllis—my Phyllis! was in the room and in my arms! Over the next five minutes, gentle reader, we will draw a curtain with your kind permission. If you have ever met your sweetheart after an absence of several months, you will readily understand why!

When we had become rational again I led her to a sofa, and, seating myself beside her, asked if her father had in any way relented towards me. At this she looked very unhappy, and for a moment I thought was going to burst into tears.

'Why! What is the matter, Phyllis, my darling?' I cried in sincere alarm. 'What is troubling you?'

'Oh, I am so unhappy,' she replied. 'Dick, there is a gentleman in Sydney now to whom papa has taken an enormous fancy, and he is exerting all his influence over me to induce me to marry him.'

'The deuce he is, and pray who may——' but I got no further in my enquiries, for at that moment I caught the sound of a footstep in the hall, and next moment Mr Wetherell opened the door. He remained for a brief period looking from one to the other of us without speaking, then he advanced, saying, 'Mr Hatteras, please be so good as to tell me when this persecution will cease? Am I not even to be free of you in my own house. Flesh and blood won't stand it, I tell you, sir— won't stand it! You pursued my daughter to England in a most ungentlemanly fashion, and now you have followed her out here again.'

'Just as I shall continue to follow her all my life, Mr Wetherell,' I replied warmly, 'wherever you may take her. I told you on board the *Orizaba*, months ago, that I loved her; well, I love her ten thousand times more now. She loves me—won't you hear her tell you so? Why then should you endeavour to keep us apart?'

'Because an alliance with you, sir, is distasteful to me in every possible way. I have other views for my daughter, you must learn.'

Here Phyllis could keep silence no longer, and broke in with—

'If you mean by that that you will force me into this hateful marriage with a man I despise, papa, you are mistaken. I will marry no one but Mr Hatteras, and so I warn you.'

We reach Australia, and the Result

'Silence, Miss! How dare you adopt that tone with me! You will do as I wish in this and all other matters, and so we'll have no more talk about it. Now, Mr Hatteras, you have heard what I have to say, and I warn you that, if you persist in this conduct, I'll see if something can't be found in the law to put a stop to it. Meanwhile, if you show yourself in my grounds again, I'll have my servants throw you out into the street! Good-day.'

Unjust as his conduct was to me, there was nothing for it but to submit, so picking up my hat I bade poor little frightened Phyllis farewell and went towards the door. But before taking my departure I was determined to have one final shot at her irascible parent, so I said, 'Mr Wetherell, I have warned you before, and I do so again: your daughter loves me, and, come what may, I will make her my wife. She is her own mistress, and you cannot force her into marrying any one against her will. Neither can you prevent her marrying me if she wishes it. You will be sorry some day that you have behaved like this to me.'

But the only answer he vouchsafed was a stormy one.

'Leave my house this instant. Not another word, sir, or I'll call my servants to my assistance!'

The stately old butler opened the front door for me, and assuming as dignified an air as was possible, I went down the drive and passed out into the street.

When I reached home again Beckenham was out, for which I was not sorry, as I wanted to have a good quiet think by myself. So, lighting a cigar, I pulled a chair into the verandah and fell to work. But I could make nothing of the situation, save that, by my interview this morning, my position with the father was, if possible, rendered even more hopeless than before. Who was this more fortunate suitor? Would it be any use my going to him and—but no, that was clearly impossible. Could I induce Phyllis to run away with me? That was possible, of course, but I rather doubted if she would care to take such an extreme step until every other means had proved unsuccessful. Then what was to be done? I began to wish that Beckenham would return in order that we might consult together.

Half an hour later our lunch was ready, but still no sign came of the youth. Where could he have got to? I waited an hour and then fell to work. Three o'clock arrived and still no sign—four, five, and even six. By this time I was in a fever of anxiety. I remembered the existence of the man who had followed us from Melbourne, and Beckenham's trusting good nature. Then and there I resolved, if he did not return before half-past seven, to set off for the nearest police station and have a search made for him. Slowly the large hand of the clock went round, and when, at the time stated, he had not appeared, I donned my hat and, enquiring the way, set off for the home of the law.

On arriving there and stating my business I was immediately conducted to the inspector in charge, who questioned me very closely as to Beckenham's appearance, age, profession, etc. Having done this, he said:

'But what reason have you, sir, for supposing that the young man has been done away with? He has only been absent from his abode, according to your statement, about eight or nine hours.'

'Simply because,' I answered, 'I have the best of reasons for knowing that ever since his arrival in Australia he has been shadowed. This morning he said he would only go for a short stroll before lunch, and I am positively certain, knowing my anxiety about him, he would not have remained away so long of his own accord without communicating with me.'

'Is there any motive you can assign for this shadowing?'

'My friend is heir to an enormous property in England. Perhaps that may assist you in discovering one?'

'Very possibly. But still I am inclined to think you are a little hasty in coming to so terrible a conclusion, Mr——?'

'Hatteras is my name, and I am staying at the "General Officer" Hotel in Palgrave Street.'

'Well, Mr Hatteras, if I were you I would go back to your hotel. You will probably find your friend there eating his dinner and thinking about instituting a search for you. If, however, he has not turned up, and does not do so by tomorrow morning, call here again and report the matter, and I will give you every assistance in my power.'

Thanking him for his courtesy I left the station and walked quickly back to the hotel, hoping to find Beckenham safely returned and at his dinner. But when the landlady met me in the verandah, and asked if I had any news of my friend, I realized that a disappointment was in store for me. By this time the excitement and worry were getting too much for me. What with Nikola, the spy, Beckenham, Phyllis, the unknown lover, and old Mr Wetherell, I had more than enough to keep my brain occupied. I sat down on a chair on the verandah with a sigh and reviewed the whole case. Nine o'clock struck by the time my reverie was finished. Just as I did so a newspaper boy came down the street lustily crying his wares. To divert my mind from its unpleasant thoughts, I called him up and bought an *Evening Mercury*. Having done so I passed into my sitting-room to read it. The first, second, and third pages held nothing of much interest to me, but on the fourth was an item which was astonishing enough to almost make my hair stand on end. It ran as follows:

RUMOURED IMPORTANT ENGAGEMENT IN HIGH LIFE.

We have it on the very best authority that an engagement will shortly be announced between a certain illustrious young nobleman, now a visitor in our city, and the beautiful daughter of one of Sydney's most prominent politicians, who has lately returned from a visit to England. The *Evening Mercury* tenders the young couple their sincerest congratulations.

Could this be the solution of the whole mystery? Could it be that the engagement of Baxter, the telegram, the idea of travel, the drugging, the imprisonment in Port Said, the substitution of the false marquis, were all means to this end? Was it possible that this man, who was masquerading as a man of title, was to marry Phyllis (for there could be no possible doubt as to the persons to whom that paragraph referred)? The very thought of such a thing was not to be endured.

There must be no delay now, I told myself, in revealing all I knew. The villains must be unmasked this very night. Wetherell should know all as soon as I could tell him.

As I came to this conclusion I crushed my paper into my pocket and set off, without a moment's delay, for Potts Point. The night was dark, and now a thick drizzle was falling.

Though it really did not take me very long, it seemed an eternity before I reached the house and rang the bell. The butler opened the door, and was evidently surprised to see me.

'Is Mr Wetherell at home?' I asked.

For a moment he looked doubtful as to what he should say, then compromising matters, answered that he would see.

'I know what that means,' I said in reply. 'Mr Wetherell is in, but you don't think he'll see me. But he must! I have news for him of the very utmost importance. Will you tell him that?'

He left me and went along the hall and upstairs. Presently he returned, shaking his head.

'I'm very sorry, sir, but Mr Wetherell's answer is, if you have anything to tell him you must put it in writing; he cannot see you.'

'But he must! In this case I can accept no refusal. Tell him, will you, that the matter upon which I wish to speak to him has nothing whatsoever to do with the request I made to him this morning. I pledge him my word on that.'

Again the butler departed, and once more I was left to cool my heels in the portico. When he returned it was with a smile upon his face.

'Mr Wetherell will be glad if you will step this way, sir.'

I followed him along the hall and up the massive stone staircase. Arriving at the top he opened a door on the left-hand side and announced 'Mr Hatteras.'

I found Mr Wetherell seated in a low chair opposite the fire, and from the fact that his right foot was resting on a sort of small trestle, I argued that he was suffering from an attack of his old enemy the gout.

'Be good enough to take a chair, Mr Hatteras,' he said, when the door had been closed. 'I must own I am quite at a loss to understand what you can have to tell me of so much importance as to bring you to my house at this time of night.'

'I think I shall be able to satisfy you on that score, Mr Wetherell,' I replied, taking the *Evening Mercury* from my pocket and smoothing it out. 'In the first place will you be good enough to tell me if there is any truth in the inference contained in that paragraph?'

I handed the paper to him and pointed to the lines in question. Having put on his glasses he examined it carefully.

'I am sorry they should have made it public so soon, I must admit,' he said. 'But I don't deny that there is a considerable amount of truth in what that paragraph reports.'

'You mean by that that you intend to try and marry Phyllis—Miss Wetherell—to the Marquis of Beckenham?'

'The young man has paid her a very considerable amount of attention ever since he arrived in the colony, and only last week he did me the honour of confiding his views to me. You see I am candid with you.'

'I thank you for it. I, too, will be candid with you. Mr Wetherell, you may set your mind at rest at once, this marriage will never take place!'

'And pray be so good as to tell me your reason for such a statement!'

'If you want it bluntly, because the young man now staying at Government House is no more the Marquis of Beckenham than I am. He is a fraud, an impostor, a cheat of the first water, put up to play his part by one of the cleverest scoundrels unhung.'

'Mr Hatteras, this is really going too far. I can quite understand your being jealous of his lordship, but I cannot understand your having the audacity to bring such a foolish charge against him. I, for one, must decline to listen to it. If he had been the fraud you make him out, how would his tutor have got those letters from his Grace the Duke of Glenbarth? Do you imagine his Excellency the Governor, who has known the family all his life, would not have discovered him ere this? No, no, sir! It won't do! If you think so, who has schooled him so cleverly? Who has pulled the strings so wonderfully?'

'Why, Nikola to be sure!'

Had I clapped a revolver to the old gentleman's head, or had the walls opened and Nikola himself stepped into the room, a greater effect of terror and consternation could not have been produced in the old gentleman's face than did those five simple words. He fell back in his chair gasping for breath, his complexion became ashen in its pallor, and for a moment his whole nervous system seemed unstrung. I sprang to his assistance, thinking he was going to have a fit, but he waved me off, and when he had recovered himself sufficiently to speak, said hoarsely:

'What do you know of Dr Nikola? Tell me for God's sake!—what do you know of him? Quick, quick!'

Thereupon I set to work and told him my story, from the day of my arrival in Sydney from Thursday Island up to the moment of my reaching his house, described my meeting and acquaintance with the real Beckenham, and all the events consequent upon it. He listened, with an awful terror growing in his face, and when I had finished my narrative with the disappearance of my friend he nearly choked.

'Mr Hatteras,' he gasped, 'will you swear this is the truth you are telling me?'

'I solemnly swear it,' I answered. 'And will do so in public when and where you please.'

'Then before I do anything else I will beg your pardon for my conduct to you. You have taken a noble revenge. I cannot thank you sufficiently. But there is not a moment to lose. My daughter is at a ball at Government House at the present moment. I should have accompanied her, but my gout would not permit me. Will you oblige me by ringing that bell?'

I rang the bell as requested, and then asked what he intended doing.

'Going off to his Excellency at once, gout or no gout, and telling him what you have told me. If it is as you have said, we must catch these scoundrels and rescue your friend without an instant's delay!'

Here the butler appeared at the door.

'Tell Jenkins to put the grey mare in my brougham and bring her round at once.'

Half an hour later we were at Government House waiting in his Excellency's study for an interview. The music of the orchestra in the ballroom came faintly in to us, and when Lord Amberley entered the room he seemed surprised, as well he might be, to see us. But as soon as he had heard what we had to tell him his expression changed.

'Mr Wetherell, this is a very terrible charge you bring against my guest. Do you think it can possibly be true?'

'I sadly fear so,' said Mr Wetherell. 'But perhaps Mr Hatteras will tell you the story exactly as he told it to me.'

I did so, and, when I had finished, the Governor went to the door and called a servant.

'Find Lord Beckenham, Johnson, at once, and ask him to be so good as to come to me here. Stay—on second thoughts I'll go and look for him myself.'

He went off, leaving us alone again to listen to the ticking of the clock upon the mantelpiece, and to wonder what was going to happen next. Five minutes went by and then ten, but still he did not return. When he did do so it was with a still more serious countenance.

'You are evidently right, gentlemen. Neither the spurious marquis, nor his tutor, Mr Baxter, can be found anywhere. I have discovered, too, that all their valuables and light luggage have been smuggled out of the house tonight without the knowledge of my servants. This is a very terrible business. But I have given instructions, and the police will be communicated with at once. Now we must do our best to find the real Beckenham.'

'Lord Amberley,' said Wetherell, in a choking voice, 'do you think one of your servants could tell my daughter to come to me at once; I am not feeling very well.'

The Governor hesitated a moment, and then said:

'I am sorry to say, Mr Wetherell, your daughter left the House an hour ago. A message was brought to her that you had been suddenly taken ill and needed her. She went off at once.'

Wetherell's anxiety was piteous to see.

'My God!' he cried in despair. 'If that is so, I am ruined. This is Nikola's revenge.'

Then he uttered a curious little sigh, moved a step forward, and fell in a dead faint upon the floor.

CHAPTER II

On the Trail

As soon as Wetherell was able to speak again he said as feebly as an old man of ninety, 'Take me home, Mr Hatteras, take me home, and let us think out together there what is best to be done to rescue my poor child.'

The Governor rose to his feet and gave him his arm.

'I think you're right, Mr Wetherell,' he said. 'It is of course just probable that you will find your daughter at her home when you arrive. God grant she may be! But in case she is not I will communicate all I know to the Police Commissioner on his arrival, and send him and his officers on to you. We must lose no time if we wish to catch these scoundrels.' Then turning to me, he continued: 'Mr Hatteras, it is owing to your promptness that we are able to take such early steps. I shall depend upon your further assistance in this matter.'

'You may do so with perfect confidence, my lord,' I answered. 'If you knew all you would understand that I am more anxious perhaps than anyone to discover the whereabouts of the young lady and my unfortunate friend.'

If his Excellency thought anything he did not give utterance to it, and Mr Wetherell's carriage being at the door we went out to it without another word. As we stepped into it Mr Wetherell cried to the coachman:

'Home, and as fast as you can go.'

Next moment we were being whirled down the drive at a pace which at any other time I should have thought dangerous. Throughout the journey we sat almost silent, wrapped in our anxieties and forebodings; hoping almost against hope that when we arrived at Potts Point we should find Phyllis awaiting us there. At last we turned into the grounds, and on reaching the house I sprang out and rang the bell, then I went down to help my companion to alight. The butler opened the door and descended the steps to take the rugs. Wetherell stopped him almost angrily, crying:

'Where is your mistress? Has she come home?'

The expression of surprise on the man's face told me, before he had time to utter a word, that our hopes were not to be realized.

'Miss Phyllis, sir?' the man said. 'Why, she's at the ball at Government 'Ouse.'

Wetherell turned from him with a deep sigh, and taking my arm went heavily up the steps into the hall.

'Come to my study, Mr Hatteras,' he said, 'and let me confer with you. For God's sake don't desert me in my hour of need!'

'You need have no fear of that,' I answered. 'If it is bad for you, think what it is for me.' And then we went upstairs together.

Reaching his study, Mr Wetherell led the way in and sat down. On a side table I noticed a decanter of whisky and some glasses. Without asking permission I went across to them and poured out a stiff nobbler for him.

'Drink this,' I said; 'it will pull you together a little; remember you will want all your strength for the work that lies before us.'

Like a child he did as he was ordered, and then sank back into his chair. I went across to the hearthrug and stood before him.

'Now,' I said, 'we must think this out from the very beginning, and to do that properly we must consider every detail. Have you any objection to answering my questions?'

'Ask any questions you like,' he replied, 'and I will answer them.'

'In the first place, then, how soon after his arrival in the colony did your daughter get to know this sham Beckenham?'

'Three days,' he answered.

'At a dance, dinner party, picnic, or what?'

'At none of these things. The young man, it appears, had seen my daughter in the street, and having been struck with her beauty asked one of the aides-de-camp at Government House, with whom we are on intimate terms, to bring him to call. At the time, I remember, I thought it a particularly friendly action on his part.'

'I don't doubt it,' I answered. 'Well that, I think, should tell us one thing.'

'And what is that?'

'That his instructions were to get to know your daughter without delay.'

'But what could his reason have been, do you think?'

'Ah, that I cannot tell you just yet. Now you must pardon what I am going to say: do you think he was serious in his intentions regarding Phyllis—I mean your daughter?'

'Perfectly, as far as I could tell. His desire, he said, was, if she would have him, to be allowed to marry her on his twenty-first birthday, which would be next week, and in proof of permission he showed me a cablegram from his father.'

'A forgery, I don't doubt. Well, then, the only construction I can put upon it is that the arrival of the real Beckenham in Sydney must have frightened him, thus compelling the gang to resort to other means of obtaining possession of her at once. Now our next business must be to find out how that dastardly act was accomplished. May I ring the bell and have up the coachman who drove your daughter to the ball?'

'By all means. Please act in every way in this matter as if this house were your own.'

I rang the bell, and when the butler appeared to answer it Mr Wetherell instructed him to find the man I wanted and send him up. The servant left the room again, and for five minutes we awaited his re-appearance in silence. When he did come back he said, 'Thompson has not come home yet, sir.'

'Not come home yet! Why, it's nearly eleven o'clock! Send him in directly he arrives. Hark! What bell is that?'

'Front door, sir.'

'Go down and answer it then, and if it should be the Commissioner of Police show him up here at once.'

As it turned out it was not the Commissioner of Police, but an Inspector.

'Good evening,' said Mr Wetherell. 'You have come from Government House, I presume?'

'Exactly so, sir,' replied the Inspector. 'His Excellency gave us some particulars and then sent us on to you.'

'You know the nature of the case?'

'His Excellency informed us himself.'

'And what steps have you taken?'

'Well, sir, to begin with, we have given orders for a thorough search throughout the city and suburbs for the tutor and the sham nobleman, at the same time more men are out looking for the real Lord Beckenham. We are also trying to find your coachman, who was supposed to have driven Miss Wetherell away from Government House, and also the carriage, which is certain to be found before very long.'

He had hardly finished speaking before there was another loud ring at the bell, and presently the butler entered the room once more. Crossing to Mr Wetherell, he said:

'Two policemen are at the front door, and they have brought Thompson home, sir.'

'Ah! We are likely to have a little light thrown upon the matter now. Let them bring him up here instantly.'

'He's not in a very nice state, sir.'

'Never mind that. Let them bring him up here, instantly!'

Again the butler departed, and a few moments later heavy footsteps ascended the stairs and approached the study door. Then two stalwart policemen entered the room supporting between them a miserable figure in coachman's livery. His hat and coat were gone and his breeches were stained with mud, while a large bruise totally obscured his left eye. His master surveyed him with unmitigated disgust.

'Stand him over there opposite me,' said Mr Wetherell, pointing to the side of the room furthest from the door.

The policemen did as they were ordered, while the man looked more dead than alive.

'Now, Thompson,' said Wetherell, looking sternly at him, 'what have you got to say for yourself?'

But the man only groaned. Seeing that in his present state he could say nothing, I went across to the table and mixed him a glass of grog. When I gave it to him he drank it eagerly. It seemed to sharpen his wits, for he answered instantly:

'It wasn't my fault, sir. If I'd only ha' known what their game was I'd have been killed afore I'd have let them do anything to hurt the young lady. But they was too cunnin' for me, sir.'

'Be more explicit, sir!' said Wetherell sternly. 'Don't stand there whining, but tell your story straightforwardly and at once.'

The poor wretch pulled himself together and did his best.

'It was in this way, sir,' he began. 'Last week I was introduced by a friend of mine to as nice a spoken man as ever I saw. He was from England, he said, and having a little money thought he'd like to try his 'and at a bit o' racing in Australia, like. He was on the look-out for a smart man, he said, who'd be able to put him up to a wrinkle or two, and maybe train for him later on. He went on to say that he'd 'eard a lot about me, and thought I was just the man for his money. Well, we got more and more friendly till the other night, Monday, when he said as how he'd settled on a farm a bit out in the country, and was going to sign the agreement, as he called it, for to rent it next day. He was goin' to start a stud farm and trainin' establishment combined, and would I take the billet of manager at three 'undred a year? Anyway, as he said, "Don't be in a 'urry to decide; take your time and think it over. Meet me at the Canary Bird 'Otel on Thursday night (that's tonight, sir) and give me your decision." Well, sir, I drove Miss Wetherell to Government 'Ouse, sir, according to orders, and then, comin' 'ome, went round by the Canary Bird to give 'im my answer, thinkin' no 'arm could ever come of it. When I drove up he was standin' at the door smoking his cigar, an' bein' an affable sort of fellow, invited me inside to take a drink. "I don't like to leave the box," I said. "Oh, never mind your horse," says he. "'Ere's a man as will stand by it for five minutes." He gave a respectable lookin' chap, alongside the lamp-post, a sixpence, and he 'eld the 'orse, so in I went. When we got inside I was for goin' to the bar, but 'e says, "No. This is an important business matter, and we don't want to be over'eard." With that he leads the way into a private room at the end of the passage and shuts the door. "What's yours?" says he. "A nobbler o' rum," says I. Then he orders a nobbler of rum for me and a nobbler of whisky for 'imself. And when it was brought we sat talkin' of the place he'd thought o' takin' an' the 'orses he was goin' to buy, an' then 'e says, "'Ullo'! Somebody listenin' at the door. I 'eard a step. Jump up and look." I got up and ran to the door, but there was nobody there, so I sat down again and we went on talking. Then he says, takin' up his glass: "'Ere's to your 'ealth, Mr Thompson, and success to the farm." We both drank it an' went on talkin' till I felt that sleepy I didn't know what to do. Then I dropped off, an' after that I don't remember nothin' of what 'appened till I woke up in the Domain, without my hat and coat, and found a policeman shakin' me by the shoulder.'

'The whole thing is as plain as daylight,' cried Wetherell bitterly. 'It is a thoroughly organized conspiracy, having me for its victim. Oh, my girlie! My poor little girlie! What has my obstinacy brought you to!'

Seeing the old man in this state very nearly broke me down, but I mastered myself with an effort and addressed a question to the unfortunate coachman:

'Pull yourself together, Thompson, and try and tell me as correctly as you can what this friend of yours was like.'

I fully expected to hear him give an exact description of the man who had followed us from Melbourne, but I was mistaken.

'I don't know, sir,' said Thompson, 'as I could rightly tell you, my mind being still a bit dizzy-like. He was tall, but not by any manner of means big made; he had very small 'ands an' feet, a sort o' what they call death's-'ead complexion; 'is 'air was black as soot, an' so was 'is eyes, an' they sparkled like two diamonds in 'is 'ead.'

'Do you remember noticing if he had a curious gold ring on his little finger, like a snake?'

'He had, sir, with two eyes made of some black stone. That's just as true as you're born.'

'Then it was Nikola,' I cried in an outburst of astonishment, 'and he followed us to Australia after all!'

Wetherell gave a deep sigh that was more like a groan than anything else; then he became suddenly a new man.

'Mr Inspector,' he cried to the police officer, 'that man, or traces of him, must be found before daylight. I know him, and he is as slippery as an eel; if you lose a minute he'll be through your fingers.'

'One moment first,' I cried. 'Tell me this, Thompson: when you drove up to the Canary Bird Hotel where did you say this man was standing?'

'In the verandah, sir.'

'Had he his hat on?'

'Yes, sir.'

'And then you went towards the bar, but it was crowded, so he took you to a private room?'

'Yes, sir.'

'And once there he began giving you the details of this farm he proposed starting. Did he work out any figures on paper?'

'Yes, sir.'

'On what?'

'On a letter or envelope; I'm not certain which.'

'Which of course he took from his pocket?'

'Yes, sir.'

'Very good,' I said. Then turning to the police officer, 'Now, Mr Inspector, shall we be off to the Canary Bird?'

'If you wish it, sir. In the meantime I'll send instructions back by these men to the different stations. Before breakfast time we must have the man who held the horse in our hands.'

'You don't know him, I suppose?' I asked Thompson.

'No, sir; but I've seen him before,' he answered.

'He's a Sydney fellow then?'

'Oh yes, sir.'

'Then there should be no difficulty in catching him. Now let us be going.'

Mr Wetherell rose to accompany us, but hard though it was to stop him I eventually succeeded in dissuading him from such a course.

'But you will let me know directly you discover anything, won't you, Mr Hatteras?' he cried as we were about to leave the room. 'Think what my anxiety will be.'

I gave my promise and then, accompanied by the Inspector, left the house. Hailing a passing cab we jumped into it and told the driver to proceed as fast as he could to the hotel in question. Just as we started a clock in the neighbourhood struck twelve. Phyllis had been in Nikola's hands three hours.

Pulling up opposite the Canary Bird (the place where the coachman had been drugged), we jumped out and bade the cabman wait. The hotel was in complete darkness, and it was not until we had pealed the bell twice that we succeeded in producing any sign of life. Then the landlord, half dressed, carrying a candle in his hand, came downstairs and called out to know who was there and what we wanted. My companion immediately said 'Police', and in answer to that magic word the door was unbarred.

'Good-evening, Mr Bartrell,' said the Inspector politely. 'May we come in for a moment on business?'

'Certainly, Mr Inspector,' said the landlord, who evidently knew my companion. 'But isn't this rather a late hour for a call. I hope there is nothing the matter?'

'Nothing much,' returned the Inspector; 'only we want to make a few enquiries about a man who was here tonight, and for whom we are looking.'

'If that is so I'm afraid I must call my barman. I was not in the bar this evening. If you'll excuse me I'll go and bring him down. In the meantime make yourselves comfortable.'

He left us to kick our heels in the hall while he went upstairs again. In about ten minutes, and just as my all-consuming impatience was well-nigh getting the better of me, he returned, bringing with him the sleepy barman.

'These gentlemen want some information about a man who was here tonight,' the landlord said by way of introduction. 'Perhaps you can give it?'

'What was he like, sir?' asked the barman of the Inspector.

The latter, however, turned to me.

'Tall, slim, with a sallow complexion,' I said, 'black hair and very dark restless eyes. He came in here with the Hon. Sylvester Wetherell's coachman.'

The man seemed to recollect him at once.

'I remember him,' he said. 'They sat in No. 5 down the passage there, and the man you mention ordered a nobbler of rum for his friend and a whisky for himself.'

'That's the fellow we want,' said the Inspector. 'Now tell me this, have you ever seen him in here before?'

'Never once,' said the barman, 'and that's a solemn fact, because if I had I couldn't have forgotten it. His figurehead wouldn't let you do that. No, sir, tonight was the first night he's ever been in the Canary Bird.'

'Did any one else visit them while they were in the room together?'

'Not as I know of. But stay, I'm not so certain. Yes; I remember seeing a tall, good-looking chap come down the passage and go in there. But it was some time, half an hour maybe, after I took in the drinks.'

'Did you see him come out again?'

'No. But I know the coachman got very drunk, and had to be carried out to the carriage.'

'How do you know that?'

'Because I saw the other two doing it.'

The Inspector turned to me.

'Not very satisfactory, is it?'

'No,' I answered. 'But do you mind letting us look into No. 5—the room they occupied?'

'Not at all,' said the landlord. 'Will you come with me?'

So saying he led the way down the passage to a little room on the right-hand side, the door of which he threw open with a theatrical flourish. It was in pitch darkness, but a few seconds later the gas was lit and we could see all that it contained. A small table stood in the centre of the room and round the walls were ranged two or three wooden chairs. A small window was at the further end and a fireplace opposite the door. On the table was a half-smoked cigar and a torn copy of the *Evening Mercury*. But that was not what I wanted, so I went down on my hands and knees and looked about upon the floor. Presently I descried a small ball of paper near the grate. Picking it up I seated myself at the table and turned to the barman, who was watching my movements attentively.

'Was this room used by any other people after the party we are looking for left?'

'No, sir. There was nobody in either of these two bottom rooms.'

'You are quite certain of that?'

'Perfectly certain.'

I took up the ball of paper, unrolled it and spread it out upon the table. To my disgust it was only the back half of an envelope, and though it had a few figures dotted about upon it, was of no possible use to us.

'Nothing there?' asked the Inspector.

'Nothing at all,' I answered bitterly, 'save a few incomprehensible figures.'

'Well, in that case, we'd better be getting up to the station and see if they've discovered anything yet.'

'Come along, then,' I answered. 'We must be quick though, for we've lost a lot of precious time, and every minute counts.'

I took up the *Evening Mercury* and followed him out to the cab, after having sincerely thanked the hotel proprietor and the barman for their courtesy. The Inspector gave the driver his orders and we set off. As we went we discussed our next movements, and while we were doing so I idly glanced at the paper I held in my hand. There was a lamp in the cab, and the light showed me on the bottom right-hand corner a round blue india-rubber stamp mark, 'W. E. Maxwell, stationer and newsagent, 23, Ipswell Street, Woolahra.'

'Stop the cab!' I almost shouted. 'Tell the man to drive us back to the Canary Bird as fast as he can go.'

The order was given, the cab faced round, and in less than a minute we were on our way back.

'What's up now?' asked the astonished Inspector.

'Only that I believe I've got a clue,' I cried.

I did not explain any further, and in five minutes we had brought the landlord downstairs again.

'I'm sorry to trouble you in this fashion,' I cried, 'but life and death depend on it. I want you to let me see No. 5 again.'

He conducted us to the room, and once more the gas was lit. The small strip of envelope lay upon the table just as I had thrown it down. I seated myself and again looked closely at it. Then I sprang to my feet.

'I thought so!' I cried excitedly, pointing to the paper; 'I told you I had a clue. Now, Mr Inspector, who wrote those figures?'

'The man you call Nikola, I suppose.'

'That's right. Now who would have bought this newspaper? You must remember that Thompson only left his box to come in here.'

'Nikola, I suppose.'

'Very good. Then according to your own showing Nikola owned this piece of envelope and this *Evening Mercury*. If that is certain, look here!'

He came round and looked over my shoulder. I pointed to what was evidently part of the gummed edge of the top of the envelope. On it were these three important words, '——swell Street, Woolahra.'

'Well,' he said, 'what about it?'

'Why, look here!' I said, as I opened the *Evening Mercury* and pointed to the stamp-mark at the bottom. 'The man who bought this newspaper at Mr Maxwell's shop also bought this envelope there. The letters "swell" before "street" constitute the last half of Ipswell, the name of the street. If that man be Nikola, as we suspect, the person who served him is certain to remember him, and it is just within the bounds of possibility he may know his address.'

'That's so,' said the Inspector, who was struck with the force of my argument. 'I know Mr Maxwell's shop, and our best plan will be to go on there as fast as we can.'

Again thanking the landlord for his civility, we returned to our cab and once more set off, this time for Mr Maxwell's shop in Ipswell Street. By the time we reached it it was nearly three o'clock, and gradually growing light.

As the cab drew up alongside the curb the Inspector jumped out and rang the bell at the side door. It was opened after awhile by a shock-headed youth, about eighteen years of age, who stared at us in sleepy astonishment.

'Does Mr Maxwell live at the shop?' asked the Inspector.

'No, sir.'

'Where then?'

'Ponson Street—third house on the left-hand side.'

'Thank you.'

Once more we jumped into the cab and rattled off. It seemed to me, so anxious and terrified was I for my darling's safety, that we were fated never to get the information we wanted; the whole thing was like some nightmare, in which, try how I would to move, every step was clogged.

A few minutes' drive brought us to Ponson Street, and we drew up at the third house on the left-hand side. It was a pretty little villa, with a nice front garden and a creeper-covered verandah. We rang the bell and waited. Presently we heard someone coming down the passage, and a moment later the door was unlocked.

'Who is there?' cried a voice from within.

'Police,' said my companion as before.

The door was immediately opened, and a very small sandy-complexioned man, dressed in a flaring suit of striped pyjamas, stood before us.

'Is anything wrong, gentlemen?' he asked nervously.

'Nothing to affect you, Mr Maxwell,' my companion replied. 'We only want a little important information, if you can give it us. We are anxious to discover a man's whereabouts before daylight, and we have been led to believe that you are the only person who can give us the necessary clue.'

'Good gracious! I never heard of such a thing. But I shall be happy to serve you if I can,' the little man answered, leading the way into his dining-room and opening the shutters with an air of importance his appearance rather belied. 'What is it?'

'Well, it's this,' I replied, producing the piece of envelope and the *Evening Mercury*. 'You see these letters on the top of this paper, don't you?' He nodded, his attention at once secured by seeing his own name. 'Well, that envelope was evidently purchased in your shop. So was this newspaper.'

'How can you tell that?'

'In the case of the envelope, by these letters; in that of the paper, by your rubber stamp on the bottom.'

'Ah! Well, now, and in what way can I help you?'

'We want to know the address of the man who bought them.'

'That will surely be difficult. Can you give me any idea of what he was like?'

'Tall, slightly foreign in appearance, distinctly handsome, sallow complexion, very dark eyes, black hair, small hands and feet.'

As my description progressed the little man's face brightened. Then he cried with evident triumph—

'I know the man; he came into the shop yesterday afternoon.'

'And his address is?'

His face fell again. His information was not quite as helpful as he had expected it would be.

'There I can't help you, I'm sorry to say. He bought a packet of paper and envelopes and the *Evening Mercury* and then left the shop. I was so struck by his appearance that I went to the door and watched him cross the road.'

'And in which direction did he go?'

'Over to Podgers' chemist shop across the way. That was the last I saw of him.'

'I'm obliged to you, Mr Maxwell,' I said, shaking him by the hand. 'But I'm sorry you can't tell us something more definite about him.' Then turning to the Inspector: 'I suppose we had better go off and find Podgers. But if we have to spend much more time in rushing about like this we shall be certain to lose them altogether.'

'Let us be off to Podgers', then, as fast as we can go.'

Bidding Mr Maxwell goodbye, we set off again, and in ten minutes had arrived at the shop and had Mr Podgers downstairs. We explained our errand as briefly as possible, and gave a minute description of the man we wanted.

'I remember him perfectly,' said the sedate Podgers. 'He came into my shop last night and purchased a bottle of chloroform.'

'You made him sign the poison book, of course?'

'Naturally I did, Mr Inspector. Would you like to see his signature?'

'Very much,' we both answered at once, and the book was accordingly produced.

Podgers ran his finger down the list.

'Brown, Williams, Davis—ah! here it is. "Chloroform: J. Venneage, 22, Calliope Street, Woolahra."'

'Venneage!' I cried. 'Why, that's not his name!'

'Very likely not,' replied Podgers; 'but it's the name he gave me.'

'Never mind, we'll try 22, Calliope Street on the chance,' said the Inspector. 'Come along, Mr Hatteras.'

Again we drove off, this time at increased pace. In less than fifteen minutes we had turned into the street we wanted, and pulled up about a hundred yards from the junction. It was a small thoroughfare, with a long line of second-class villa residences on either side. A policeman was sauntering along on the opposite side of the way, and the Inspector called him over. He saluted respectfully, and waited to be addressed.

'What do you know of number 22?' asked the Inspector briefly. The constable considered for a few moments, and then said:

'Well, to tell you the truth, sir, I didn't know until yesterday that it was occupied.'

'Have you seen anybody about there?'

'I saw three men go in just as I came on the beat tonight.'

'What were they like?'

'Well, I don't know that I looked much at them. They were all pretty big, and they seemed to be laughing and enjoying themselves.'

'Did they! Well, we must go in there and have a look at them. You had better come with us.'

We walked on down the street till we arrived at No. 22. Then opening the gate we went up the steps to the hall door. It was quite light enough by this time to enable us to see everything distinctly. The Inspector gave the bell a good pull and the peal re-echoed inside the house. But not a sound of any living being came from within in answer. Again the bell was pulled, and once more we waited patiently, but with the same result.

'Either there's nobody at home or they refuse to hear,' said the Inspector. 'Constable, you remain where you are and collar the first man you see. Mr Hatteras, we will go round to the back and try to effect an entrance from there.'

We left the front door, and finding a path reached the yard. The house was only a small one, with a little verandah at the rear on to which the back door opened. On either side of the door were two fair-sized windows, and by some good fortune it chanced that the catch of one of these was broken.

Lifting the sash up the Inspector jumped into the room, and as soon as he was through I followed him. Then we looked about us. The room, however, was destitute of furniture or occupants.

'I don't hear anybody about,' my companion said, opening the door that led into the hall. Just at that moment I heard a sound, and touching his arm signed to him to listen. We both did so, and surely enough there came again the faint muttering of a human voice. In the half-dark of the hall it sounded most uncanny.

'Somebody in one of the front rooms,' said the Inspector. 'I'll slip along and open the front door, bring in the man from outside, and then we'll burst into the room and take our chance of capturing them.'

He did as he proposed, and when the constable had joined us we moved towards the room on the left.

Again the mutterings came from the inside, and the Inspector turned the handle of the door. It was locked, however.

'Let me burst it is,' I whispered.

He nodded, and I accordingly put my shoulder against it, and bringing my strength to bear sent it flying in.

Then we rushed into the room, to find it, at first glance, empty.

Just at that moment, however, the muttering began again, and we looked towards the darkest corner; somebody was there, lying on the ground. I rushed across and knelt down to look. *It was Beckenham; his mouth gagged and his hands and feet bound. The noise we had heard was that made by him trying to call us to his assistance.*

In less time than it takes to tell I had cut his bonds and helped him to sit up. Then I explained to the Inspector who he was.

'Thank God you're found!' I cried. 'But what does it all mean? How long have you been like this? and where is Nikola?'

'I don't know how long I've been here,' he answered, 'and I don't know where Nikola is.'

'But you must know something about him!' I cried. 'For Heaven's sake tell me all you can! I'm in awful trouble, and your story may give me the means of saving a life that is dearer to me than my own.'

'Get me something to drink first, then,' he replied; 'I'm nearly dying of thirst; after that I'll tell you all I can.'

Fortunately I had had the foresight to put a flask of whisky into my pocket, and I now took it out and gave him a stiff nobbler. It revived him somewhat, and he prepared to begin his tale. But the Inspector interrupted:

'Before you commence, my lord, I must send word to the Commissioner that you have been found.'

He wrote a message on a piece of paper and dispatched the constable with it. Having done so he turned to Beckenham and said—

'Now, my lord, pray let us hear your story.'

Beckenham forthwith commenced.

CHAPTER III

Lord Beckenham's Story

'When you left me, Mr Hatteras, to visit Miss Wetherell at Potts Point, I remained in the house for half an hour or so reading. Then, thinking no harm could possibly come of it, I started out for a little excursion on my own account. It was about half-past eleven then.

'Leaving the hotel I made for the ferry and crossed Darling Harbour to Millers Point; then, setting myself for a good ramble, off I went through the city, up one street and down another, to eventually bring up in the botanical gardens. The view was so exquisite that I sat myself down on a seat and resigned myself to rapturous contemplation of it. How long I remained there I could not possibly say. I only know that while I was watching the movements of a man-o'-war in the cove below me I became aware, by intuition—for I did not look at him—that I was the object of close scrutiny by a man standing some little distance from me. Presently I found him drawing closer to me, until he came boldly up and seated himself beside me. He was a queer-looking little chap, in some ways not unlike my old tutor Baxter, with a shrewd, clean-shaven face, grey hair, bushy eyebrows, and a long and rather hooked nose. He was well dressed, and when he spoke did so with some show of education. When we had been sitting side by side for some minutes he turned to me and said,

'"It is a beautiful picture we have spread before us, is it not?"

'"It is, indeed," I answered. "And what a diversity of shipping?"

'"You may well say that," he continued. "It would be an interesting study, would it not, to make a list of all the craft that pass in and out of this harbour in a day—to put down the places where they were built and whence they hail, the characters of their owners and commanders, and their errands about the world. What a book it would make, would it not? Look at that man-o'-war in Farm Cove; think of the money she cost, think of where that money came from—the rich people who paid without thinking, the poor who dreaded the coming of the tax collector like a visit from the Evil One; imagine the busy dockyard in which she was built—can't you seem to hear the clang of the riveters and the buzzing of the steam saws? Then take that Norwegian boat passing the fort there; think of her birthplace in far Norway, think of the places she has since seen, imagine her masts growing in the forests on the mountainside of lonely fjords, where the silence is so intense that a stone rolling down and dropping into the water echoes like thunder. Then again, look at that emigrant vessel steaming slowly up the harbour; think of the folk aboard her, every one with his hopes and fears, confident of a successful future in this *terra incognita*, or despondent of that and everything else. Away to the left there you see a little island schooner making her way down towards the blue Pacific; imagine her in a few weeks among the islands—tropical heavens dropped down into sunlit waters—buying

Lord Beckenham's Story

such produce as perhaps you have never heard of. Yes, it is a wonderful picture—a very wonderful picture?"

'"You seem to have studied it very carefully," I said, after a moment's silence.

'"Perhaps I have," he answered. "I am deeply interested in the life of the sea—few more so. Are you a stranger in New South Wales?"

'"Quite a stranger," I replied. "I only arrived in Australia a few days since."

'"Indeed! Then you have to make the acquaintance of many entrancing beauties yet. Forgive my impertinence, but if you are on a tour, let me recommend you to see the islands before you return to your home."

'"The South-Sea Islands, I presume you mean?" I said.

'"Yes; the bewitching islands of the Southern Seas! The most entrancingly beautiful spots on God's beautiful earth! See them before you go. They will amply repay any trouble it may cost you to reach them."

'"I should like to see them very much," I answered, feeling my enthusiasm rising at his words.

'"Perhaps you are interested in them already," he continued.

'"Very much indeed," I replied.

'"Then, in that case, I may not be considered presumptuous if I offer to assist you. I am an old South-Sea merchant myself, and I have amassed a large collection of beautiful objects from the islands. If you would allow me the pleasure I should be delighted to show them to you."

'"I should like to see them very much indeed," I answered, thinking it extremely civil of him to make the offer.

'"If you have time we might perhaps go and overhaul them now. My house is but a short distance from the Domain, and my carriage is waiting at the gates."

'"I shall be delighted," I said, thinking there could be no possible harm in my accepting his courteous invitation.

'"But before we go, may I be allowed to introduce myself," the old gentleman said, taking a card-case from his pocket and withdrawing a card. This he handed to me, and on it I read—

Mr Mathew Draper.

' "I am afraid I have no card to offer you in return," I said; "but I am the Marquis of Beckenham."

' "Indeed! Then I am doubly honoured," the old gentleman said, with a low bow. "Now shall we wend our way up towards my carriage?"

'We did so, chatting as we went. At the gates a neat brougham was waiting for us, and in it we took our places.

' "Home," cried my host, and forthwith we set off down the street. Up one thoroughfare and down another we passed, until I lost all count of our direction. Throughout the drive my companion talked away in his best style; commented on the architecture of the houses, had many queer stories to tell of the passers-by, and in many other ways kept my attention engaged till the carriage came to a standstill before a small but pretty villa in a quiet street.

'Mr Draper immediately alighted, and when I had done so, dismissed his coachman, who drove away as we passed through the little garden and approached the dwelling. The front door was opened by a dignified man-servant, and we entered. The hall, which was a spacious one for so small a dwelling, was filled with curios and weapons, but I had small time for observing them, as my host led me towards a room at the back. As we entered it he said, "I make you welcome to my house, my lord. I hope, now that you have taken the trouble to come, I shall be able to show you something that will repay your visit." Thereupon, bidding me seat myself for a few moments, he excused himself and left the room. When he returned he began to do the honours of the apartment. First we examined a rack of Australian spears, nulla-nullas, and boomerangs, then another containing New Zealand hatchets and clubs. After this we crossed to a sort of alcove where reposed in cases a great number of curios collected from the further islands of the Pacific. I was about to take up one of these when the door on the other side of the room opened and someone entered. At first I did not look round, but hearing the newcomer approaching me I turned, to find myself, to my horrified surprise, face to face *with no less a person than Dr Nikola*. He was dressed entirely in black, his coat was buttoned and displayed all the symmetry of his peculiar figure, while his hair seemed blacker and his complexion even paler than before. He had evidently been prepared for my visit, for he held out his hand and greeted me without a sign of astonishment upon his face.

' "This is indeed a pleasure, my lord," he said, still with his hand outstretched, looking hard at me with his peculiar cat-like eyes. "I did not expect to see you again so soon. And you are evidently a little surprised at meeting me."

' "I am more than surprised," I answered bitterly, seeing how easily I had been entrapped. "I am horribly mortified and angry. Mr Draper, you had an easy victim."

'Mr Draper said nothing, but Dr Nikola dropped into a chair and spoke for him.

' "You must not blame my old friend Draper," he said suavely. "We have been wondering for the last twenty-four hours how we might best get hold of you, and the means we have employed so successfully seemed the only possible way. Have no fear, my lord, you shall not be hurt. In less than twenty-four hours you will enjoy the society of your energetic friend Mr Hatteras again."

' "What is your reason for abducting me like this?" I asked. "You are foolish to do so, for Mr Hatteras will leave no stone unturned to find me."

' "I do not doubt that at all," said Dr Nikola quietly; "but I think Mr Hatteras will find he will have all his work cut out for him this time."

' "If you imagine that your plans are not known in Sydney you are mistaken," I cried. "The farce you are playing at Government House is detected, and Mr Hatteras, directly he finds I am lost, will go to Lord Amberley and reveal everything."

' "I have not the slightest objection," returned Dr Nikola quietly. "By the time Mr Hatteras can take those steps—indeed, by the time he discovers your absence at all, we shall be beyond the reach of his vengeance."

'I could not follow his meaning, of course, but while he had been speaking I had been looking stealthily round me for a means of escape. The only way out of the room was, of course, by the door, but both Nikola and his ally were between me and that. Then a big stone hatchet hanging on the wall near me caught my eye. Hardly had I seen it before an idea flashed through my brain. Supposing I seized it and fought my way out. The door of the room stood open, and I noticed with delight that the key was in the lock on the outside. One rush, armed with the big hatchet, would take me into the passage; then before my foes could recover their wits I might be able to turn the key, and, having locked them in, make my escape from the house before I could be stopped.

'Without another thought I made up my mind, sprang to the wall, wrenched down the hatchet, and prepared for my rush. But by the time I had done it both Nikola and Draper were on their feet.

'"Out of my way!" I cried, raising my awful weapon aloft. "Stop me at your peril!"

'With my hatchet in the air I looked at Nikola. He was standing rigidly erect, with one arm outstretched, the hand pointing at me. His eyes glared like living coals, and when he spoke his voice came from between his teeth like a serpent's hiss.

'"Put down that axe!" he said.

'With that the old horrible fear of him which had seized me on board ship came over me again. His eyes fascinated me so that I could not look away from them. I put down the hatchet without another thought. Still he gazed at me in the same hideous fashion.

'"Sit down in that chair," he said quietly. "You cannot disobey me." And indeed I could not. My heart was throbbing painfully, and an awful dizziness was creeping over me. Still I could not get away from those terrible eyes. They seemed to be growing larger and fiercer every moment. Oh! I can feel the horror of them even now. As I gazed his white right hand was moving to and fro before me with regular sweeps, and with each one I felt my own will growing weaker and weaker. That I was being mesmerized, I had no doubt, but if I had been going to be murdered I could not have moved a finger to save myself.

'Then there came a sudden but imperative knock at the door, and both Nikola and Draper rose. Next moment the man whom we had noticed in the train as we came up from Melbourne, and against whom you, Mr Hatteras, had warned me in Sydney, entered the room. He crossed and stood respectfully before Nikola.

'"Well, Mr Eastover, what news?" asked the latter. "Have you done what I told you?"

'"Everything," the man answered, taking an envelope from his pocket. "Here is the letter you wanted."

'Nikola took it from his subordinate's hand, broke the seal, and having withdrawn the contents, read it carefully. All this time, seeing resistance was quite useless, I did not move. I felt too sick and giddy for anything. When he had finished his correspondence Nikola said

Lord Beckenham's Story

something in an undertone to Draper, who immediately left the room. During the time he was absent none of us spoke. Presently he returned, bringing with him a wine glass filled with water, which he presented to Nikola.

'"Thank you," said that gentleman, feeling in his waistcoat pocket. Presently he found what he wanted and produced what looked like a small silver scent-bottle. Unscrewing the top, he poured from it into the wine glass a few drops of some dark-coloured liquid. Having done this he smelt it carefully and then handed it to me.

'"I must ask you to drink this, my lord," he said. "You need have no fear of the result: it is perfectly harmless."

'Did ever man hear such a cool proposition? Very naturally I declined to do as he wished.

'"You *must* drink it!" he reiterated. "Pray do so at once. I have no time to waste bandying words with you."

'"I will not drink it!" I cried, rising to my feet, and prepared to make a fight for it if need should be.

'Once more those eyes grew terrible, and once more that hand began to make the passes before my face. Again I felt the dizziness stealing over me. His will was growing every moment too strong for me. I could not resist him. So when he once more said, "Drink!" I took the glass and did as I was ordered. After that I remember seeing Nikola, Draper, and the man they called Eastover engaged in earnest conversation on the other side of the room. I remember Nikola crossing to where I sat and gazing steadfastly into my face, and after that I recollect no more until I came to my senses in this room, to find myself bound and gagged. For what seemed like hours I lay in agony, then I heard footsteps in the verandah, and next moment the sound of voices. I tried to call for help, but could utter no words. I thought you would go away without discovering me, but fortunately for me you did not do so. Now, Mr Hatteras, I have told you everything; you know my story from the time you left me up to the present moment.'

For some time after the Marquis had concluded his strange story both the Inspector and I sat in deep thought. That Beckenham had been kidnapped in order that he should be out of the way while the villainous plot for abducting Phyllis was being enacted there could be no doubt. But why had he been chosen? and what clues were we to gather from what he had told us? I turned to the Inspector and said,

'What do you think will be the best course for us to pursue now?'

'I have been wondering myself. I think, as there is nothing to be learned from this house, the better plan would be for you two gentleman to go back to Mr Wetherell, while I return to the detective office and see if anything has been discovered by the men there. As soon as I have found out I will join you at Potts Point. What do you think?'

I agreed that it would be the best course; so, taking the Marquis by the arms (for he was still too weak to walk alone), we left the house, and were about to step into the street when I stopped, and asking them to wait for me ran back into the room again. In the corner, just as it had been thrown down, lay the rope with which Beckenham had been bound and the pad which had been fitted over his mouth. I picked both up and carried them into the verandah.

'Come here, Mr Inspector,' I cried. 'I thought I should learn something from this. Take a look at this rope and this pad, and tell me what you make of them.'

He took each up in turn and looked them over and over. But he only shook his head.

'I don't see anything to guide us,' he said as he laid them down again.

'Don't you?' I cried. 'Why, they tell me more than I have learnt from anything else I've seen. Look at the two ends of this.' (Here I took up the rope and showed it to him.) 'They're seized!'

I looked triumphantly at him, but he only stared at me in surprise, and said, 'What do you mean by "seized"?'

'Why, I mean that the ends are bound up in this way—look for yourself. Now not one landsman in a hundred *seizes* a rope's end. This line was taken from some ship in the harbour, and—— By Jove! here's another discovery!'

'What now?' he cried, being by this time almost as excited as I was myself.

'Why, look here,' I said, holding the middle of the rope up to the light, so that we could get a better view of it. 'Not very many hours ago this rope was running through a block, and that block was rather an uncommon one.'

'How do you know that it was an uncommon one?'

'Because it has been newly painted, and what's funnier still, painted green, of all other colours. Look at this streak of paint along the line; see how it's smudged. Now let's review the case as we walk along.'

So saying, with the Marquis between us, we set off down the street, hoping to be able to pick up an early cab.

'First and foremost,' I said, 'remember old Draper's talk of the South Seas—remember the collection of curios he possessed. Probably he owns a schooner, and it's more than probable that this line and this bit of canvas came from it.'

'I see what you're driving at,' said the Inspector. 'It's worth considering. Directly I get to the office I will set men to work to try and find this mysterious gentleman. You would know him again, my lord?'

'I should know him anywhere,' was Beckenham's immediate reply.

'And have you any idea at all where this house, to which he conducted you, is located?'

'None at all. I only know that it was about half-way down a street of which all the houses, save the one at the corner—which was a grocer's shop—were one-storeyed villas.'

'Nothing a little more definite, I suppose?'

'Stay! I remember that there was an empty house with broken windows almost opposite, and that on either side of the steps leading up to the front door were two stone eagles with outstretched wings. The head of one of the eagles—the left, I think—was missing.'

The Inspector noted these things in his pocket-book, and just as he had finished we picked up a cab and called it to the sidewalk. When we had got in and given the driver Mr Wetherell's address, I said to the Inspector:

'What are you going to do first?'

'Put some men on to find Mr Draper, and some more to find an island schooner with her blocks newly painted green.'

'You won't be long in letting us know what you discover, will you?' I said. 'Remember how anxious we are.'

'You may count on my coming to you at once with any news I may procure,' he answered.

A few moments later we drew up at Mr Wetherell's door. Bidding the Inspector goodbye we went up the steps and rang the bell. By the time the cab was out in the street again we were in the house making our way, behind the butler, to Mr Wetherell's study.

The old gentleman had not gone to bed, but sat just as I had left him so many hours before. As soon as we were announced he rose to receive us.

'Thank God, Mr Hatteras, you have come back!' he said. 'I have been in a perfect fever waiting for you. What have you to report?'

'Not very much, I'm afraid,' I answered. 'But first let me have the pleasure of introducing the real Marquis of Beckenham to you, whom we have had the good fortune to find and rescue.'

Mr Wetherell bowed gravely and held out his hand.

'My lord,' he said, 'I am thankful that you have been discovered. I look upon it as one step towards the recovery of my poor girl. I hope now that both you and Mr Hatteras will take up your abode with me during the remainder of your stay in the colony. You have had a scurvy welcome to New South Wales. We must see if we can't make up to you for it. But you look thoroughly worn out; I expect you would like to go to bed.'

He rang the bell, and when his butler appeared, gave him some instructions about preparing rooms for us.

Ten minutes later the man returned and stated that our rooms were ready, whereupon Mr Wetherell himself conducted Beckenham to the apartment assigned to him. When he returned to me, he asked if I would not like to retire too, but I would not hear of it. I could not have slept a wink, so great was my anxiety. Seeing this, he seated himself and listened attentively while I gave him an outline of Beckenham's story. I had hardly finished before I heard a carriage roll up to the door. There was a ring at the bell, and presently the butler, who, like ourselves, had not dreamt of going to bed, though his master had repeatedly urged him to do so, entered and announced the Inspector.

Wetherell hobbled across to receive him with an anxious face. 'Have you any better tidings for me?' he asked

'Not very much, I'm afraid, sir,' the Inspector said, shaking his head. 'The best I have to tell you is that your carriage and horse have been found in the yard of an empty house off Pitt Street.'

'Have you been able to discover any clue as to who put them there?'

'Not one! The horse was found out of the shafts tied to the wall. There was not a soul about the place.'

Lord Beckenham's Story

Wetherell sat down again and covered his face with his hands. At that instant the telephone bell in the corner of the room rang sharply. I jumped up and went across to it. Placing the receivers to my ears, I heard a small voice say, 'Is that Mr Wetherell's house, Potts Point?'

'Yes,' I answered.

'Who is speaking?'

'Mr Hatteras. Mr Wetherell, however, is in the room. Who are you?'

'Detective officer. Will you tell Mr Wetherell that Mr Draper's house has been discovered?'

I communicated the message to Mr Wetherell, and then the Inspector joined me at the instrument and spoke.

'Where is the house?' he enquired.

'83, Charlemagne Street—north side.'

'Very good. Inspector Murdkin speaking. Let plain clothes men be stationed at either end of the street, and tell them to be on the look out for Draper, and to wait for me. I'll start for the house at once.'

'Very good, sir.'

He rang off and then turned to me.

'Are you too tired to come with me, Mr Hatteras?' he enquired.

'Of course not,' I answered. 'Let us go at once.'

'God bless you!' said Wetherell. 'I hope you may catch the fellow.'

Bidding him goodbye, we went downstairs again, and jumped into the cab, which was directed to the street in question.

Though it was a good distance from our starting-point, in less than half an hour we had pulled up at the corner. As the cab stopped, a tall man, dressed in blue serge, who had been standing near the lamp-post, came forward and touched his hat.

'Good-morning, Williams,' said the Inspector. 'Any sign of our man?'

'Not one, sir. He hasn't come down the street since I've been here.'

'Very good. Now come along and we'll pay the house a visit.'

So saying he told the cabman to follow us slowly, and we proceeded down the street. About half-way along he stopped and pointed to a house on the opposite side.

'That is the house his lordship mentioned, with the broken windows, and this is where Mr Draper dwells, if I am not much mistaken—see the eagles are on either side of the steps, just as described.'

It was exactly as Beckenham had told us, even to the extent of the headless eagle on the left of the walk. It was a pretty little place, and evidently still occupied, as a maid was busily engaged cleaning the steps.

Pushing open the gate, the Inspector entered the little garden and accosted the girl.

'Good-morning,' he said politely. 'Pray, is your master at home?'

'Yes, sir; he's at breakfast just now.'

'Well, would you mind telling him that two gentlemen would like to see him?'

'Yes, sir.'

The girl rose to her feet, and, wiping her hands on her apron, led the way into the house. We followed close behind her. Then, asking us to wait a moment where we were, she knocked at a door on the right and opening it, disappeared within.

'Now,' said the Inspector, 'our man will probably appear, and we shall have him nicely.'

The Inspector had scarcely spoken before the door opened again, and a man came out. To our surprise, however, he was very tall and stout, with a round, jovial face, and a decided air of being satisfied with himself and the world in general.

'To what do I owe the honour of this visit?' he said, looking at the Inspector.

'I am an Inspector of Police, as you see,' answered my companion, 'and we are looking for a man named Draper, who yesterday was in possession of this house.'

'I am afraid you have made some little mistake,' returned the other. 'I am the occupier of this house, and have been for some months past. No Mr Draper has anything at all to do with it.'

The Inspector's face was a study for perfect bewilderment. Nor could mine have been much behind it. The Marquis had given such a minute description of the dwelling opposite and the two stone birds on the steps, that there could be no room for doubt that this *was* the house. And yet it was physically impossible that this man could be Draper; and, if it were the place where Beckenham had been drugged, why were the weapons, etc., he had described not in the hall?

'I cannot understand it at all,' said the Inspector, turning to me. 'This is the house, and yet where are the things with which it ought to be furnished?'

'You have a description of the furniture, then?' said the owner. 'That is good, for it will enable me to prove to you even more clearly that you are mistaken. Pray come and see my sitting-rooms for yourselves.'

He led the way into the apartment from which he had been summoned, and we followed him. It was small and nicely furnished, but not a South-Sea curio or native weapon was to be seen in it. Then we followed him to the corresponding room at the back of the house. This was upholstered in the latest fashion; but again there was no sign of what Beckenham had led us to expect we should find. We were completely nonplussed.

'I am afraid we have troubled you without cause,' said the Inspector, as we passed out into the hall again.

'Don't mention it,' the owner answered; 'I find my compensation in the knowledge that I am not involved in any police unpleasantness.'

'By the way,' said the Inspector suddenly, 'have you any idea who your neighbours may be?'

'Oh, dear, yes!' the man replied. 'On my right I have a frigidly respectable widow of Low Church tendencies. On my left, the Chief Teller of the Bank of New Holland. Both very worthy members of society, and not at all the sort of people to be criminally inclined.'

'In that case we can only apologize for our intrusion and wish you good-morning.'

'Pray don't apologize. I should have been glad to have assisted you. Good-morning.'

We went down the steps again and out into the street. As we passed through the gate, the Inspector stopped and examined a mark on the right hand post. Then he stooped and picked up what looked like a pebble. Having done so we resumed our walk.

'What on earth can be the meaning of it all?' I asked. 'Can his lordship have made a mistake?'

'No, I think not. We have been cleverly duped, that's all.'

'What makes you think so?'

'I didn't think so until we passed through the gate on our way out. Now I'm certain of it. Come across the street.'

I followed him across the road to a small plain-looking house, with a neatly-curtained bow window and a brass plate on the front door. From the latter I discovered that the proprietress of the place was a dressmaker, but I was completely at a loss to understand why we were visiting her.

As soon as the door was opened the Inspector asked if Miss Tiffins were at home, and, on being told that she was, enquired if we might see her. The maid went away to find out, and presently returned and begged us to follow her. We did so down a small passage towards the door of the room which contained the bow window.

Miss Tiffins was a lady of uncertain age, with a prim, precise manner, and corkscrew curls. She seemed at a loss to understand our errand, but bade us be seated, and then asked in what way she could be of service to us.

'In the first place, madam,' said the Inspector, 'let me tell you that I am an officer of police. A serious crime has been perpetrated, and I have reason to believe that it may be in your power to give us a clue to the persons who committed it.'

'You frighten me, sir,' replied the lady. 'I cannot at all see in what way I can help you. I lead a life of the greatest quietness. How, therefore, can I know anything of such people?'

'I do not wish to imply that you do know anything of them. I only want you to carry your memory back as far as yesterday, and to answer me the few simple questions I may ask you.'

'I will answer them to the best of my ability.'

'Well, in the first place, may I ask if you remember seeing a brougham drive up to that house opposite about mid-day yesterday?'

'No, I cannot say that I do,' the old lady replied after a moment's consideration.

'Do you remember seeing a number of men leave the house during the afternoon?'

'No. If they came out I did not notice them.'

'Now, think for one moment, if you please, and tell me what vehicles, if any, you remember seeing stop there.'

'Let me try to remember. There was Judge's baker's cart, about three, the milk about five, and a furniture van about half-past six.'

'That's just what I want to know. And have you any recollection whose furniture van it was?'

'Yes. I remember reading the name as it turned round. Goddard & James, George Street. I wondered if the tenant was going to move.'

The Inspector rose, and I followed his example.

Lord Beckenham's Story

'I am exceedingly obliged to you, Miss Tiffins. You have helped me materially.'

'I am glad of that,' she answered; 'but I trust I shall not be wanted to give evidence in court. I really could not do it.'

'You need have no fear on that score,' the Inspector answered. 'Good-day.'

'Good-day.'

When we had left the house the Inspector turned to me and said:

'It was a great piece of luck finding a dressmaker opposite. Commend me to ladies of that profession for knowing what goes on in the street. Now we will visit Messrs Goddard & James and see who hired the things. Meantime, Williams' (here he called the plain-clothes constable to him), 'you had better remain here and watch that house. If the man we saw comes out, follow him, and let me know where he goes.'

'Very good, sir,' the constable replied, and we left him to his vigil.

Then, hailing a passing cab, we jumped into it and directed the driver to convey us to George Street. By this time it was getting on for midday, and we were both worn out. But I was in such a nervous state that I could not remain inactive. Phyllis had been in Nikola's hands nearly fourteen hours, and so far we had not obtained one single definite piece of information as to her whereabouts.

Arriving at the shop of Messrs Goddard & James, we went inside and asked to see the chief partner. An assistant immediately conveyed us to an office at the rear of the building, where we found an elderly gentleman writing at a desk. He looked up as we entered, and then, seeing the Inspector's uniform, rose and asked our business.

'The day before yesterday,' began my companion, 'you supplied a gentleman with a number of South-Sea weapons and curios on hire, did you not?'

'I remember doing so—yes,' was the old gentleman's answer. 'What about it?'

'Only I should be glad if you would favour me with a description of the person who called upon you about them—or a glimpse of his letter, if he wrote.'

'He called and saw me personally.'

'Ah! That is good. Now would you be so kind as to describe him?'

'Well, in the first place, he was very tall and rather handsome; he had, if I remember rightly, a long brown moustache, and was decidedly well dressed.'

'That doesn't tell us very much, does it? Was he alone?'

'No. He had with him, when he came into the office, an individual whose face singularly enough remains fixed in my memory—indeed I cannot get it out of my head.'

Instantly I became all excitement.

'What was this second person like?' asked the Inspector.

'Well, I can hardly tell you—that is to say, I can hardly give you a good enough description of him to make you see him as I saw him. He was tall and yet very slim, had black hair, a sallow complexion, and the blackest eyes I ever saw in a man. He was clean-shaven and exquisitely dressed, and when he spoke, his teeth glittered like so many pearls. I never saw another man like him in my life.'

'Nikola for a thousand!' I cried, bringing my hand down with a thump upon the table.

'It looks as if we're on the track at last,' said the Inspector. Then, turning to Mr Goddard again: 'And may I ask now what excuse they made to you for wanting these things!'

'They did not offer any; they simply paid a certain sum down for the hire of them, gave me their address, and then left.'

'And the address was?'

'83, Charlemagne Street. Our van took the things there and fetched them away last night.'

'Thank you. And now one or two other questions. What name did the hirer give?'

'Eastover.'

'And when they left your shop how did they go away?'

'A cab was waiting at the door for them, and I walked out to it with them.'

'There were only two of them, you think?'

'No. There was a third person waiting for them in the cab, and it was that very circumstance which made me anxious to have my things brought back as soon as possible. If I had been able to, I should have even declined to let them go.'

'Why so?'

'Well, to tell you that would involve a story. But perhaps I had better tell you. It was in this way. About three years ago, through a distant relative, I got to know a man named Draper.'

'Draper!' I cried. 'You don't mean—but there, I beg your pardon. Pray go on.'

'As I say, I got to know this man Draper, who was a South-Sea trader. We met once or twice, and then grew more intimate. So friendly did we at last become, that I even went so far as to put some money into a scheme he proposed to me. It was a total failure. Draper proved a perfect fraud and a most unbusiness-like person, and all I got out of the transaction was the cases of curios and weapons which this man Eastover hired from me. It was because—when I went out with my customers to their cab—I saw this man Draper waiting for them that I became uneasy about my things. However, all's well that ends well, and as they returned my goods and paid the hire I must not grumble.'

'And now tell me what you know of Draper's present life,' the Inspector said.

'Ah! I'm afraid of that I can tell you but little. He has been twice declared bankrupt, and the last time there was some fuss made over his schooner, the *Merry Duchess*.'

'He possesses a schooner, then?'

'Oh, yes! A nice boat, She's in harbour now, I fancy.'

'Thank you very much, Mr Goddard. I am obliged to you for your assistance in this matter.'

'Don't mention it. I hope that what I have told you may prove of service to you.'

'I'm sure it will. Good-day.'

'Good-day, gentlemen.'

He accompanied us to the door, and then bade us farewell.

'Now what are we to do?' I asked.

'Well, first, I am going back to the office to put a man on to find this schooner, and then I'm going to take an hour or two's rest. By that time we shall know enough to be able to lay our hands on Dr Nikola and his victim, I hope.'

'God grant we may!'

'Where are you going now?'

'Back to Potts Point,' I answered.

We thereupon bade each other farewell and set off in different directions.

When I reached Mr Wetherell's house I learned from the butler that his master had fallen asleep in the library. Not wishing to disturb him, I enquired the whereabouts of my own bedroom, and on being conducted to it, laid myself down fully dressed upon the bed. So utterly worn out was I, that my head had no sooner touched the pillow than I was fast asleep. How long I lay there I do not know, but when I woke it was to find Mr Wetherell standing beside me, holding a letter in his hand. He was white as a sheet, and trembling in every limb.

'Read this, Mr Hatteras,' he cried. 'For Heaven's sake tell me what we are to do!'

I sat up on the side of the bed and read the letter he handed to me. It was written in what was evidently a disguised hand, on common notepaper, and ran as follows:

TO MR WETHERELL,
  POTTS POINT, SYDNEY.

DEAR SIR,
  This is to inform you that your daughter is in very safe keeping. If you wish to find her you had better be quick about it. What's more, you had better give up consulting the police, and such like, in the hope of getting hold of her. The only way you *can* get her will be to act as follows: At eight o'clock tonight charter a boat and pull down the harbour as far as Shark Point. When you get there, light your pipe three times, and someone in a boat near by will do the same. Be sure to bring with you the sum of *one hundred thousand pounds in gold, and—this is most important—bring with you the little stick you got from China Pete, or do not come at all.* Above all, do not bring more than one man. If you do not put in an appearance you will not hear of your daughter again. Yours obediently,
                                    THE MAN WHO KNOWS.

CHAPTER IV

Following up a Clue

For some moments after I had perused the curious epistle Mr Wetherell had brought to my room I remained wrapped in thought.

'What do you make of it?' my companion asked.

'I don't know what to say,' I answered, looking at it again. 'One thing, however, is quite certain, and that is that, despite its curious wording, it is intended you should take it seriously.'

'You think so?'

'I do indeed. But I think when the Inspector arrives it would be just as well to show it to him. What do you say?'

'I agree with you. Let us defer consideration of it until we see him.'

When, an hour later, the Inspector put in an appearance, the letter was accordingly placed before him, and his opinion asked concerning it. He read it through without comment, carefully examined the writing and signature, and finally held it up to the light. Having done this he turned to me and said:

'Have you that envelope we found at the Canary Bird, Mr Hatteras?'

I took it out of my pocket and handed it to him. He then placed it on the table side by side with the letter, and through a magnifying-glass scrutinized both carefully. Having done so, he asked for the envelope in which it had arrived. Mr Wetherell had thrown it into the waste-paper basket, but a moment's search brought it to light. Again he scrutinized both the first envelope and the letter, and then compared them with the second cover.

'Yes, I thought so,' he said. 'This letter was written either by Nikola, or at his desire. The paper is the same as that he purchased at the stationer's shop we visited.'

'And what had we better do now?' queried Wetherell, who had been eagerly waiting for him to give his opinion.

'We must think,' said the Inspector. 'In the first place, I suppose you don't feel inclined to pay the large sum mentioned here?'

'Not if I can help it, of course,' answered Wetherell. 'But if the worst comes to the worst, and I cannot rescue my poor girl any other way, I would sacrifice even more than that.'

'Well, we'll see if we can find her without compelling you to pay anything at all,' the Inspector cried. 'I've got an idea in my head.'

'And what is that?' I cried; for I, too, had been thinking out a plan.

'Well, first and foremost,' he answered, 'I want you, Mr Wetherell, to tell me all you can about your servants. Let us begin with the butler. How long has he been with you?'

'Nearly twenty years.'

'A good servant, I presume, and a trustworthy man?'

'To the last degree. I have implicit confidence in him.'

'Then we may dismiss him from our minds. I think I saw a footman in the hall. How long has he been with you?'

'Just about three months.'

'And what sort of a fellow is he?'

'I really could not tell you very much about him. He seems intelligent, quick and willing, and up to his work.'

'Is your cook a man or a woman?'

'A woman. She has been with me since before my wife's death—that is to say, nearly ten years. You need have no suspicion of her.'

'Housemaids?'

'Two. Both have been with me some time, and seem steady, respectable girls. There is also a kitchen-maid; but she has been with me nearly as long as my cook, and I would stake my reputation on her integrity.'

'Well, in that case, the only person who seems at all suspicious is the footman. May we have him up?'

'With pleasure. I'll ring for him.'

Mr Wetherell rang the bell, and a moment later it was answered by the man himself.

'Come in, James, and shut the door behind you,' his master said.

The man did as he was ordered, but not without looking, as I thought, a little uncomfortable. The Inspector I could see had noticed this too, for he had been watching him intently ever since he had appeared in the room.

'James,' said Mr Wetherell, 'the Inspector of Police wishes to ask you a few questions. Answer him to the best of your ability.'

'To begin with,' said the Inspector, 'I want you to look at this envelope. Have you seen it before?'

He handed him the envelope of the anonymous letter addressed to Mr Wetherell. The man took it and turned it over in his hands.

'Yes, sir,' he said, 'I have seen it before; I took it in at the front door.'

'From whom?'

'From a little old woman, sir,' the man answered.

'A little old woman!' cried the Inspector, evidently surprised. 'What sort of woman?'

'Well, sir, I don't know that I can give you much of a description of her. She was very small, had a sort of nut-cracker face, a little black poke bonnet, and walked with a stick.'

'Should you know her again if you saw her?'

'Oh yes, sir.'

'Did she say anything when she gave you the letter?'

'Only, "For Mr Wetherell, young man." That was all, sir.'

'And you didn't ask if there was an answer? That was rather a singular omission on your part, was it not?'

'She didn't give me time, sir. She just put it into my hand and went down the steps again.'

'That will do. Now, Mr Wetherell, I think we'd better see about getting that money from the bank. You need not wait, my man.'

The footman thereupon left the room, while both Mr Wetherell and I stared at the Inspector in complete astonishment. He laughed.

'You are wondering why I said that,' he remarked at last.

'I must confess it struck me as curious,' answered Wetherell.

'Well, let me tell you I did it with a purpose. Did you notice that young man's face when he entered the room and when I gave him the letter? There can be no doubt about it, he is in the secret.'

'You mean that he is in Nikola's employ? Then why don't you arrest him?'

'Because I want to be quite certain first. I said that about the money because, if he is Nikola's agent, he will carry the information to him, and by so doing keep your daughter in Sydney for at least a day longer. Do you see?'

'I do, and I admire your diplomacy. Now what is your plan?'

'May I first tell mine?' I said.

'Do,' said the Inspector, 'for mine is not quite matured yet.'

'Well,' I said, 'my idea is this. I propose that Mr Wetherell shall obtain from his bank a number of gold bags, fill them with lead discs to represent coin, and let it leak out before this man that he has got the money in the house. Then tonight Mr Wetherell will set off for the waterside. I will row him down the harbour disguised as a boatman. We will pick up the boat, as arranged in that letter. In the meantime you must start from the other side in a police boat, pull up to meet us, and arrest the man. Then we will force him to disclose Miss Wetherell's whereabouts, and act upon his information. What do you say?'

'It certainly sounds feasible,' said the Inspector, and Mr Wetherell nodded his head approvingly. At that moment the Marquis entered the room, looking much better than when we had found him on the preceding night, and the conversation branched off into a different channel.

My plot seemed to commend itself so much to Mr Wetherell's judgement, that he ordered his carriage and drove off there and then to his bank, while I went down to the harbour, arranged about a boat, and having done so, proceeded up to the town, where I purchased a false beard, an old dungaree suit, such as a man loafing about the harbour might wear, and a slouch hat of villainous appearance. By the time I got back to the house Mr Wetherell had returned. With great delight he conducted me to his study, and, opening his safe, showed me a number of canvas bags, on each of which was printed £1,000.

'But surely there are not £100,000 there?'

'No,' said the old gentleman with a chuckle. 'There is the counterfeit of £50,000 there; for the rest I propose to show them these.'

So saying, he dived his hand into a drawer and produced a sheaf of crisp banknotes.

'There—these are notes for the balance of the amount.'

'But you surely are not going to pay? I thought we were going to try to catch the rascals without letting any money change hands.'

'So we are; do not be afraid. If you will only glance at these notes you will see that they are dummies, every one of them. They are for me to exhibit to the man in the boat; in the dark they'll pass muster, never fear.'

'Very good indeed,' I said with a laugh. 'By the time they can be properly examined we shall have the police at hand ready to capture him.'

'I believe we shall,' the old gentleman cried, rubbing his hands together in his delight—'I believe we shall. And a nice example we'll make of the rascals. Nikola thinks he can beat me; I'll show him how mistaken he is!'

And for some time the old gentleman continued in this strain, confidently believing that he would have his daughter with him again by the time morning came. Nor was I far behind him in confidence. Since Nikola had not spirited her out of the country my plot seemed the one of all others to enable us to regain possession of her; and not only that, but we hoped it would give us an opportunity of punishing those who had so schemed against her. Suddenly an idea was born in my brain, and instantly I acted on it.

'Mr Wetherell,' I said, 'supposing, when your daughter is safe with you again, I presume so far as once more to offer myself for your son-in-law, what will you say?'

'What will I say?' he cried. 'Why, I will tell you that you shall have her, my boy, with ten thousand blessings on your head. I know you now; and since I've treated you so badly, and you've taken such a noble revenge, why, I'll make it up to you, or my name's not Wetherell. But we won't talk any more about that till we have got possession of her; we have other and more important things to think of. What time ought we to start tonight?'

'The letter fixes the meeting for ten o'clock; we had better be in the boat by half-past nine. In the meantime I should advise you to take a little rest. By the way, do you think your footman realizes that you have the money?'

'He ought to, for he carried it up to this room for me; and, what's more, he has applied for a holiday this afternoon.'

'That's to carry the information. Very good; everything is working excellently. Now I'm off to rest for a little while.'

'I'll follow your example. In the meantime I'll give orders for an early dinner.'

We dined at seven o'clock sharp, and at half-past eight I went off to my room to don my disguise; then, bidding the Marquis goodbye—much to the young gentleman's disgust, for he was most anxious to accompany us—I slipped quietly out of my window, crossed the garden—I hoped unobserved—and then went down to the harbour side, where the boat I had chartered was waiting for me. A quarter of an hour later Wetherell's carriage drove up, and on seeing it I went across and opened the door. My disguise was so perfect that for a moment the old gentleman seemed undecided whether to trust me or not. But my voice, when I spoke, reassured him, and then we set to work carrying the bags of spurious money down to the boat. As soon as this was accomplished we stepped in. I seated myself amidships and got out the oars, Mr Wetherell taking the yoke-lines in the stern. Then we shoved off, and made our way out into the harbour.

It was a dull, cloudy night, with hardly a sign of a star in the whole length and breadth of heaven, while every few minutes a cold, cheerless wind swept across the water. So chilly indeed was it that before we had gone very far I began to wish I had added an overcoat to my other disguises. We hardly spoke, but pulled slowly down towards the island mentioned in the letter. The strain on our nerves was intense, and I must confess to feeling decidedly nervous as I wondered what would happen if the police boat did not pull up to meet us, as we had that morning arranged.

A quarter to ten chimed from some church ashore as we approached within a hundred yards of our destination. Then I rested on my oars and waited. All round us were the lights of bigger craft, but no rowing-boat could I see. About five minutes before the hour I whispered to Wetherell to make ready, and in answer the old gentleman took a matchbox from his pocket. Exactly as the town clocks struck the hour he lit a vesta; it flared a little and then went out. As it did so a boat shot out of the darkness to port. He struck a second, and then a third. As the last one burned up and then died away, the man rowing the boat I have just referred to struck a light, then another, then another, in rapid succession. Having finished his display, he took up his oars and propelled his boat towards us. When he was within talking distance he said in a gruff voice:

'Is Mr Wetherell aboard?'

To this my companion immediately answered, not however without a tremble in his voice, 'Yes, here I am!'

'Money all right?'

'Can you see if I hold it up?' asked Mr Wetherell. As he spoke a long black boat came into view on the other side of our questioner, and pulled slowly towards him. It was the police boat.

'No, I don't want to see,' said the voice again. 'But this is the message I was to give you. Pull in towards Circular Quay and find the *Maid of the Mist* barque. Go aboard her, and take your money down into the cuddy. There you'll get your answer.'

'Nothing more?' cried Mr Wetherell.

'That's all I was told,' answered the man, and then said, 'Goodnight.'

At the same moment the police boat pulled up alongside him and made fast. I saw a dark figure enter his boat, and next moment the glare of a lantern fell upon the man's face. I picked up my oars and pulled over to them, getting there just in time to hear the Inspector ask the man his name.

'James Burbidge,' was the reply. 'I don't know as how you've got anything against me. I'm a licensed waterman, I am.'

'Very likely,' said the Inspector; 'but I want a little explanation from you. How do you come to be mixed up in this business?'

'What—about this 'ere message, d'you mean?'

'Yes, about this message. Where is it from? Who gave it to you?'

'Well, if you'll let me go, I'll tell you all about it,' growled the man. 'I was up at the Hen and Chickens this evenin', just afore dark, takin' a nobbler along with a friend. Presently in comes a cove in a cloak. He beckons me outside and says, "Do you want to earn a sufring?"—a sufring is twenty bob. So I says, "My word, I do!" Then he says, "Well, you go out on the harbour tonight, and be down agin Shark Point at ten?" I said I would, and so I was. "You'll see a boat there with an old gent in it," says he. "He'll strike three matches, and you do the same. Then ask him if he's Mr Wetherell. If he says 'Yes', ask him if the money's all right? And if he says 'Yes' to that, tell him to pull in towards Circular Quay and find the *Maid of the Mist* barque. He's to take his money down to the cuddy, and he'll get his answer there." That's the truth, so 'elp me bob! I don't know what you wants to go arrestin' of an honest man for.'

The Inspector turned to the water police.

'Does any man here know James Burbidge?'

Two or three voices immediately answered in the affirmative, and this seemed to decide the officer, for he turned to the waterman again and said, 'As some of my men seem to know you, I'll let you off. But for your own sake go home and keep a silent tongue in your head.'

He thereupon clambered back into his own boat and bade the man depart. In less time than it takes to tell he was out of sight. We then drew up alongside the police boat.

'What had we better do, Mr Inspector?' asked Mr Wetherell.

'Find the *Maid of the Mist* at once. She's an untenanted ship, being for sale. You will go aboard, sir, with your companion, and down to the cuddy. Don't take your money, however. We'll draw up alongside as soon as you're below, and when one of their gang, whom you'll dispatch for it, comes up to get the coin, we'll collar him, and then come to your assistance. Do you understand?'

'Perfectly. But how are we to know the vessel?'

'Well, the better plan would be for you to follow us. We'll pull to within a hundred yards of her. I learn from one of my men here that she's painted white, so you'll have no difficulty in recognising her.'

'Very well, then, go on, and we'll follow you.'

The police boat accordingly set off, and we followed about fifty yards behind her. A thick drizzle was now falling, and it was by no means an easy matter to keep her in sight. For some time we pulled on. Presently we began to get closer to her. In a quarter of an hour we were alongside.

'There's your craft,' said the Inspector, pointing as he spoke to a big vessel showing dimly through the scud to starboard of us. 'Pull over to her.'

I followed his instructions, and, arriving at the vessel's side, hitched on, made the painter fast, and then, having clambered aboard, assisted Mr Wetherell to do the same. As soon as we had both gained the deck we stood and looked about us, at the same time listening for any sound which might proclaim the presence of the men we had come to meet; but save the sighing of the wind in the shrouds overhead, the dismal creaking of blocks, and the drip of moisture upon the deck, no sign was to be heard. There was nothing for it, therefore, but to make our way below as best we could. Fortunately I had had the forethought to bring with me a small piece of candle, which came in very handily at the present juncture, seeing that the cuddy, when we reached the companion ladder, was wrapt in total darkness. Very carefully I stepped inside, lit the candle, and then, with Mr Wetherell at my heels, made my way down the steps.

Arriving at the bottom we found ourselves in a fair-sized saloon of the old-fashioned type. Three cabins stood on either side, while from the bottom of the companion ladder, by which we had descended, to a long cushioned locker right aft under the wheel, ran a table covered with American cloth. But there was no man of any kind to be seen. I opened cabin after cabin, and searched each with a like result. We were evidently quite alone in the ship.

'What do you make of it all?' I asked of Mr Wetherell.

'It looks extremely suspicious,' he answered. 'Perhaps we're too early for them. But see, Mr Hatteras, there's something on the table at the further end.'

So there was—something that looked very much like a letter. Together we went round to the end of the table, and there, surely enough, found a letter pinned to the American cloth, and addressed to my companion in a bold but rather quaint handwriting.

'It's for you, Mr Wetherell,' I said, removing the pins and presenting it to him. Thereupon we sat down beside the table, and he broke the seal with trembling fingers. It was not a very long epistle, and ran as follows:

My Dear Mr Wetherell,

Bags of imitation money and spurious banknotes will not avail you, nor is it politic to arrange that the Water Police should meet you on the harbour for the purpose of arresting me. You have lost your opportunity, and your daughter accordingly leaves Australia tonight. I will, however, give you one more chance—take care that you make the most of it. The sum I now ask is £150,000, *with the stick given you by China Pete*, and must be paid without enquiry of any sort. If you are agreeable to this, advertise as follows, 'I will Pay—W., and give stick!' in the agony column *Sydney Morning Herald*, on the 18th, 19th, and 20th of this present month. Further arrangements will then be made with you.

The Man who Knows.

'Oh, my God, I've ruined all!' cried Mr Wetherell as he put the letter down on the table; 'and, who knows? I may have killed my poor child!'

Seeing his misery, I did my best to comfort him; but it was no use. He seemed utterly broken down by the failure of our scheme, and, if the truth must be told, my own heart was quite as heavy. One thing was very certain, there was a traitor in our camp. Someone had overheard our plans and carried them elsewhere. Could it be the footman? If so, he should have it made hot for him when I got sufficient proof against him; I could promise him that most certainly. While I was thinking over this, I heard a footstep on the companion stairs, and a moment later the Inspector made his appearance. His astonishment at finding us alone, reading a letter by the light of one solitary candle, was unmistakable, for he said, as he came towards us and sat down, 'Why, how's this? Where are the men?'

'There are none. We've been nicely sold,' I answered, handing him the letter. He perused it without further remark, and when he had done so, sat drumming with his fingers upon the table in thought.

'We shall have to look in your own house for the person who has given us away, Mr Wetherell!' he said at last. 'The folk who are running this affair are as cute as men are made nowadays; it's a pleasure to measure swords with them.'

'What do you think our next move had better be?'

'Get home as fast as we can. I'll return with you, and we'll talk it over there. It's no use our remaining here.'

We accordingly went on deck, and descended to our wherry again. This time the Inspector accompanied us, while the police boat set off down the harbour on other business. When we had seen it pull out into the darkness, we threw the imitation money overboard, pushed off for the shore, landed where we had first embarked, and then walked up to Mr Wetherell's house. It was considerably after twelve o'clock by the time we reached it, but the butler was still sitting up for us. His disappointment seemed as keen as ours when he discovered that we had returned without his young mistress. He followed us up to the study with spirits and glasses, and then at his master's instruction went off to bed.

'Now, gentlemen,' began Mr Wetherell, when the door had closed upon him, 'let us discuss the matter thoroughly. But before we begin, may I offer you cigars.'

The Inspector took one, but I declined, stating that I preferred a pipe. But my pipe was in my bedroom, which was on the other side of the passage; so asking them to wait for me, I went to fetch it. I left the room, shutting the door behind me. But it so happened that the pipe-case had been moved, and it was some minutes before I could find it. Having done so, however, I blew out my candle, and was about to leave the room, which was exactly opposite the study, when I heard the green baize door at the end of the passage open, and a light footstep come along the corridor. Instantly I stood perfectly still, and waited to see who it might be. Closer and closer the step came, till I saw in the half dark the pretty figure of one of the parlour maids. On tiptoe she crept up to the study door, and then stooping down, listened at the keyhole. Instantly I was on the alert, every nerve strained to watch her. For nearly five minutes she stood there, and then with a glance round, tiptoed quietly along the passage again, closing the baize door after her.

When she was safely out of hearing I crossed to the study. Both the Inspector and Mr Wetherell saw that something had happened, and were going to question me. But I held up my hand.

'Don't ask any questions, but tell me as quickly, and as nearly as you can, what you have been talking about during the last five minutes,' I said.

'Why?'

'Don't stop to ask questions. Believe in the importance of my haste. What was it?'

'I have only been giving Mr Wetherell a notion of the steps I propose to take,' said the Inspector.

'Thank you. Now I'm off. Don't sit up for me, Mr Wetherell; I'm going to follow up a clue that may put us on the right scent at last. I don't think you had better come, Mr Inspector, but I'll meet you here again at six o'clock.'

'You can't explain, I suppose?' said the latter, looking a little huffed.

'I'm afraid not,' I answered; 'but I'll tell you this much—I saw one of the female servants listening at this door just now. She'll be off, if I mistake not, with the news she has picked up, and I want to watch her. Good-night.'

'Good-night, and good luck to you.'

Without another word I slipped off my boots, and carrying them in my hand, left the room, and went downstairs to the morning-room. This apartment looked out over the garden, and possessed a window shaded by a big tree. Opening it, I jumped out and carefully closed it after me. Then, pausing for a moment to resume my boots, I crept quietly down the path, jumped a low wall, and so passed into the back street. About fifty yards from the tradesmen's entrance, but on the opposite side of the road, there was a big Moreton Bay fig-tree. Under this I took my stand, and turned a watchful eye upon the house. Fortunately it was a dark night, so that it would have been extremely difficult for any one across the way to have detected my presence.

For some minutes I waited, and was beginning to wonder if I could have been deceived, when I heard the soft click of a latch, and next moment a small dark figure passed out into the street, and closed the gate after it. Then, pausing a moment as if to make up her mind, for the mysterious person was a woman, she set off quickly in the direction of the city. I followed about a hundred yards behind her.

With the exception of one policeman, who stared very hard at me, we did not meet a soul. Once or twice I nearly lost her, and when we reached the city itself I began to see that it would be well for me to decrease the difference that separated us, if I did not wish to bid goodbye to her altogether. I accordingly hastened my steps, and in this fashion we passed up one street and down another, until we reached what I cannot help thinking must have been the lowest quarter of Sydney. On either hand were Chinese names and sign-boards, marine stores, slop shops, with pawnbrokers and public-houses galore; while in this locality few of the inhabitants seemed to have any idea of what bed meant. Groups of sullen-looking men and women were clustered at the corners, and on one occasion the person I was pursuing was stopped by them. But she evidently knew how to take care of herself, for she was soon marching on her way again.

At the end of one long and filthily dirty street she paused and looked about her. I had crossed the road just before this, and was scarcely ten yards behind her. Pulling my hat well down to shade my face, and sticking my hands in my pockets, I staggered and reeled along, doing my best to imitate the gait of a drunken man. Seeing only me about, she went up to the window of a corner house and tapped with her knuckles thrice upon the glass. Before one could have counted twenty the door of the dwelling was opened, and she passed in. Now I was in a nasty fix—either I must be content to abandon my errand, or I must get inside the building, and trust to luck to procure the information I wanted. Fortunately, in my present disguise the girl would be hardly likely to recognize her master's guest. So giving them time to get into a room, I also went up to the door and turned the handle. To my delight it was unlocked. I opened it, and entered the house.

The passage was in total darkness; but I could make out where the door of the room I wanted to find was located by a thin streak of light low down upon the floor. As softly as I possibly could, I crept up to it, and bent down to look through the keyhole. The view was necessarily limited, but I could just make out the girl I had followed sitting upon a bed; while leaning against the wall, a dirty clay pipe in her mouth, was the vilest old woman I have ever in my life set eyes on. She was very small, with a pinched-up nut-cracker face, dressed in an old bit of tawdry finery, more than three sizes too large for her. Her hair fell upon her shoulders in a tangled mass, and from under it her eyes gleamed out like those of a wicked little Scotch terrier ready to bite. As I bent down to listen I heard her say:

'Well, my pretty dear, and what information have you got for the gentleman, that brings you down at this time of night?'

'Only that the *coppers* are going to start at daylight looking for the *Merry Duchess*. I heard the Inspector say so himself.'

'At daylight, are they?' croaked the old hag. 'Well, I wish 'em joy of their search, I do—them—them! Any more news, my dear?'

'The master and that long-legged slab of a Hatteras went out tonight down the harbour. The old man brought home a lot of money bags, but what was in 'em was only dummies.'

'I know that, too, my dear. Nicely they was sold. Ha! ha!'

She chuckled like an old fiend, and then began to cut up another pipe of tobacco in the palm of her hand like a man. She smoked negro head, and the reek of it came out through the keyhole to me. But the younger woman was evidently impatient, for she rose and said:

'When do they sail with the girl, Sally?'

'They're gone, my dear. They went at ten tonight.'

At this piece of news my heart began to throb painfully, so much indeed that I could hardly listen for its beating.

'They weren't long about it,' said the younger girl critically.

'That Nikola's not long about anything,' remarked the old woman.

'I hope Pipa Lannu will agree with her health—the stuck-up minx—I do!' the younger remarked spitefully. 'Now where's the money he said I was to have. Give it to me and let me be off. I shall get the sack if this is found out.'

'It was five pound I was to give yer, wasn't it?' the elder woman said, pushing her hand deep down into her pocket.

'Ten,' said the younger sharply. 'No larks, Sally. I know too much for you!'

'Oh, you know a lot, honey, don't you? Of course you'd be expected to know more than old Aunt Sally, who's never seen anything at all, wouldn't you? Go along with you!'

'Hand me over the money I say, and let me be off!'

'Of course you do know a lot more, don't you? There's a pound!'

While they were wrangling over the payment I crept down the passage again to the front door. Once I had reached it, I opened it softly and went out, closing it carefully behind me. Then I took to my heels and ran down the street in the direction I had come. Enquiring my way here and there from policemen, I eventually reached home, scaled the wall, and went across the garden to the morning-room window. This I opened, and by its help made my way into the house and upstairs. As I had expected that he would have gone to bed, my astonishment was considerable at meeting Mr Wetherell on the landing.

'Well, what have you discovered?' he asked anxiously as I came up to him.

'Information of the greatest importance,' I answered; 'but one other thing first. Call up your housekeeper, and tell her you have reason to believe that one of the maids is not in the house. Warn her not to mention you in the matter, but to discharge the girl before breakfast. By the time you've done that I'll have changed my things and be ready to tell you everything.'

'I'll go and rouse her at once I'm all impatience to know what you have discovered.'

He left me and passed through the green baize door to the servants' wing; while I went to my bedroom and changed my things. This done, I passed into the study, where I found a meal awaiting me. To this I did ample justice, for my long walk and the excitement of the evening had given me an unusual appetite.

Just as I was cutting myself a third slice of beef Mr Wetherell returned, and informed me that the housekeeper was on the alert, and would receive the girl on her reappearance.

'Now tell me of your doings,' said old gentleman.

I thereupon narrated all that had occurred since I left the study in search of my pipe—how I had seen the girl listening at the door, how I had followed her into the town; gave him a description of old Sally, the maid's interview with her, and my subsequent return home. He listened eagerly, and, when I had finished, said:

'Do you believe then that my poor girl has been carried off by Nikola to this island called Pipa Lannu?'

'I do; there seems to be no doubt at all about it.'

'Well then, what are we to do to rescue her? Shall I ask the Government to send a gunboat down?'

'If you think it best; but, for my own part, I must own I should act independently of them. You don't want to make a big sensation, I presume; and remember, to arrest Nikola would be to open the whole affair.'

'Then what do you propose?'

'I propose,' I answered, 'that we charter a small schooner, fit her out, select half a dozen trustworthy and silent men, and then take our departure for Pipa Lannu. I am well acquainted with the island, and, what's more, I hold a master's certificate. We would sail in after dark, arm all our party thoroughly, and go ashore. I expect they will be keeping your daughter a prisoner in a hut. If that is so, we will surround it and rescue her without any trouble or fuss, and, what is better still, without any public scandal. What do you think?'

'I quite agree with what you say. I think it's an excellent idea; and, while you've been speaking, I too have been thinking of something. There's my old friend McMurtough, who has a nice steam yacht. I'm sure he'd be willing to let us have the use of her for a few weeks.'

'Where does he live?—far from here?'

'His office would be best; we'll go over and see him directly after breakfast if you like.'

'By all means. Now I think I'll go and take a little nap; I feel quite worn out. When the Inspector arrives you will be able to explain all that has happened; but I think I should ask him to keep a quiet tongue in his head about the island. If it leaks out at all, it may warn them, and they'll be off elsewhere—to a place perhaps where we may not be able to find them.'

'I'll remember,' said Mr Wetherell, and thereupon I retired to my room, and, having partially undressed, threw myself upon my bed. In less than two minutes I was fast asleep, never waking until the first gong sounded for breakfast; then, after a good bath, which refreshed me wonderfully, I dressed in my usual habiliments, and went downstairs. Mr Wetherell and the Marquis were in the dining-room, and when I entered both he and the Marquis, who held a copy of the *Sydney Morning Herald* in his hand, seemed prodigiously excited.

'I say, Mr Hatteras,' said the latter (after I had said 'Good-morning'), 'here's an advertisement which is evidently intended for you!'

'What is it about?' I asked. 'Who wants to advertise for me?'

'Read for yourself,' said the Marquis, giving me the paper.

I took it, and glanced down the column to which he referred me until I came to the following:

Richard Hatteras.—If this should meet the eye of Mr Richard Hatteras, of Thursday Island, Torres Straits, lately returned from England, and believed to be now in Sydney; he is earnestly requested to call at the office of Messrs Dawson & Gladman, Solicitors, Castlereagh Street, where he will hear of something to his advantage.

There could be no doubt at all that I was the person referred to; but what could be the reason of it all? What was there that I could possibly hear to my advantage, save news of Phyllis, and it would be most unlikely that I would learn anything about the movements of the gang who had abducted her from a firm of first-class solicitors such as I understood Messrs Dawson & Gladman to be. However, it was no use wondering about it, so I dismissed the matter from my mind for the present, and took my place at the table. In the middle of the meal the butler left the room, in response to a ring at the front door. When he returned, it was to inform me that a man was in the hall, who wished to have a few moments' conversation with me. Asking Mr Wetherell to excuse me, I left the room.

In the hall I found a seedy-looking individual of about middle age. He bowed, and on learning that my name was Hatteras, asked if he might be permitted five minutes alone with me. In response, I led him to the morning-room, and having closed the door, pointed to a seat.

'What is your business?' I enquired, when he had sat down.

'It is rather a curious affair to approach, Mr Hatteras,' the man began. 'But to commence, may I be permitted to suggest that you are uneasy in your mind about a person who has disappeared?'

'You may certainly suggest that, if you like,' I answered cautiously.

'If it were in a man's power to furnish a clue regarding that person's whereabouts, it might be useful to you, I suppose,' he continued, craftily watching me out of the corners of his eyes.

'Very useful,' I replied. 'Are you in a position to do so?'

'I might possibly be able to afford you some slight assistance,' he went on. 'That is, of course, provided it were made worth my while.'

'What do you call "worth your while"?'

'Well, shall we say five hundred pounds? That's not a large sum for really trustworthy information. I ought to ask a thousand, considering the danger I'm running in mixing myself up with the affair. Only I'm a father myself, and that's why I do it.'

'I see. Well, let me tell you, I consider five hundred too much.'

'Well then I'm afraid we can't trade. I'm sorry.'

'So am I. But I'm not going to buy a pig in a poke.'

'Shall we say four hundred, then?'

'No. Nor three—two, or one. If your information is worth anything, I don't mind giving you fifty pounds for it. But I won't give a halfpenny more.'

As I spoke, I rose as if to terminate the interview. Instantly my visitor adopted a different tone.

'My fault is my generosity,' he said. 'It's the ruin of me. Well, you shall have it for fifty. Give me the money, and I'll tell you.'

'By no means,' I answered. 'I must hear the information first. Trust to my honour. If what you tell me is worth anything, I'll give you fifty pounds for it. Now what is it?'

'Well, sir, to begin with, you must understand that I was standing at the corner of Pitt Street an evening or two back, when two men passed me talking earnestly together. One of 'em was a tall strapping fellow, the other a little chap. I never saw two eviller looking rascals in my life. Just as they came alongside me, one says to the other, "Don't be afraid; I'll have the girl at the station all right at eight o'clock sharp." The other said something that I could not catch, and then I lost sight of them. But what I had heard stuck in my head, and so I accordingly went off to the station, arriving there a little before the hour. I hadn't been there long before the smallest of the two chaps I'd seen in the street came on to the platform, and began looking about him. By the face of him he didn't seem at all pleased at not finding the other man waiting for him. A train drew up at the platform, and presently, just before it started, I saw the other and a young lady wearing a heavy veil come quickly along. The first man saw them, and gave a little cry of delight. "I thought you'd be too late," says he. "No fear of that," says the other, and jumps into a first-class carriage, telling the girl to get in after him, which she does, crying the while, as I could see. Then the chap on the platform says to the other who was leaning out of the window, "Write to me from Bourke, and tell me how she gets on." "You bet," says his friend. "And don't you forget to keep your eye on Hatteras." "Don't you be afraid," answered the man on the platform. Then the guard whistled, and the train went out of the station. Directly I was able to I got away, and first thing this morning came on here. Now you have my information, and I'll trouble you for that fifty pound.'

'Not so fast, my friend. Your story seems very good, but I want to ask a few questions first. Had the bigger man—the man who went up to Bourke, a deep cut over his left eye?'

'Now I come to think of it, he had. I'd forgotten to tell you that.'

'So it was he, then? But are you certain it was Miss Wetherell? Remember she wore a veil. Could you see if her hair was flaxen in colour?'

'Very light it was; but I couldn't see rightly which colour it was.'

'You're sure it was a light colour?'

'Quite sure. I could swear to it in a court of law if you wanted me to.'

'That's all right then, because it shows me your story is a fabrication. Come, get out of this house or I'll throw you out. You scoundrel, for two pins I'd give you such a thrashing as you'd remember all your life!'

'None o' that, governor. Don't you try it on. Hand us over that fifty quid.'

With that the scoundrel whipped out a revolver and pointed it at me. But before he could threaten again I had got hold of his wrist with one hand, snatched the pistol with the other, and sent him sprawling on his back upon the carpet.

'Now, you brute,' I cried, 'what am I going to do with you, do you think? Get up and clear out of the house before I take my boot to you.'

He got up and began to brush his clothes.

'I want my fifty pound,' he cried.

'You'll get more than you want if you come here again,' I said. 'Out you go!'

With that I got him by the collar and dragged him out of the room across the hall, much to the butler's astonishment, through the front door, and then kicked him down the steps. He fell in a heap on the gravel.

'All right, my fine bloke,' he said as he lay there; 'you wait till I get you outside. I'll fix you up, and don't you make no mistake.'

I went back to the dining-room without paying any attention to his threats. Both Mr Wetherell and Beckenham had been witnesses of what had occurred, and now they questioned me concerning his visit. I gave them an outline of the story the man had told me and convinced them of its absurdity. Then Mr Wetherell rose to his feet.

'Now shall we go and see McMurtough?'

'Certainly,' I said; 'I'll be ready as soon as you are.'

'You will come with us I hope, Lord Beckenham?' Wetherell said.

'With every pleasure,' answered his lordship, and thereupon we went off to get ready.

Three-quarters of an hour later we were sitting in Mr McMurtough's ante-room, waiting for an interview. At the end of ten minutes a commissionaire came in to inform us that Mr McMurtough was disengaged, and forthwith conducted us to his room. We found him a small, grey-haired, pleasant-looking gentleman, full of life and fun. He received Mr Wetherell as an old friend, and then waited to be introduced to us.

'Let me make you acquainted with my friends, McMurtough,' said Wetherell—'the Marquis of Beckenham and Mr Hatteras.'

He bowed and then shook hands with us, after which we sat down and Wetherell proceeded to business. The upshot of it all was that he fell in with our plans as soon as we had uttered them, and expressed himself delighted to lend his yacht in such a good cause.

'I only wish I could come with you,' he said; 'but unfortunately that is quite impossible. However, you are more than welcome to my boat. I will give you a letter, or send one to the Captain, so that she may be prepared for sea today. Will you see about provisioning her, or shall I?'

'We will attend to that,' said Wetherell. 'All the expenses must of course be mine.'

'As you please about that, my old friend,' returned McMurtough.

'Where is she lying?' asked Wetherell.

The owner gave us the direction, and then having sincerely thanked him, we set off in search of her. She was a nice craft of about a hundred and fifty tons burden, and looked as if she ought to be a good sea boat. Chartering a wherry, we were pulled off to her. The captain was below when we arrived, but a hail brought him on deck. Mr Wetherell then explained our errand, and gave him his owner's letter. He read it through, and having done so, said:

'I am at your service, gentlemen. From what Mr McMurtough says here I gather that there is no time to lose, so with your permission I'll get to work at once.'

'Order all the coal you want, and tell the steward to do the same for anything he may require in his department. The bills must be sent in to me.'

'Very good, Mr Wetherell. And what time will you be ready?'

'As soon as you are. Can you get away by three o'clock this afternoon, think you?'

'Well, it will be a bit of a scramble, but I think we can manage it. Anyhow, I'll do my best, you may be sure of that, sir.'

'I'm sure you will. There is grave need for it. Now we'll go back and arrange a few matters ashore. My man shall bring our baggage down later on.'

'Very good, sir. I'll have your berths prepared.'

With that we descended to the boat again, and were pulled ashore. Arriving there, Mr Wetherell asked what we should do first.

'Hadn't we better go up to the town and purchase a few rifles and some ammunition?' I said. 'We can have them sent down direct to the boat, and so save time.'

'A very good suggestion. Let us go at once.'

We accordingly set off for George Street—to a shop I remembered having seen. There we purchased half a dozen Winchester repeaters, with a good supply of ammunition. They were to be sent down to the yacht without fail that morning. This done, we stood on the pavement debating what we should do next. Finally it was decided that Mr Wetherell and Beckenham should go home to pack, while I made one or two other small purchases, and then join them. Accordingly, bidding them goodbye, I went on down the street, completed my business, and was about to hail a cab and follow them, when a thought struck me: Why should I not visit Messrs Dawson & Gladman, and find out why they were advertising for me? This I determined to do, and accordingly set off for Castlereagh Street. Without much hunting about I discovered their office, and went inside.

In a small room leading off the main passage, three clerks were seated. To them I addressed myself, asking if I might see the partners.

'Mr Dawson is the only one in town, sir,' said the boy to whom I spoke. 'If you'll give me your name, I'll take it in to him.'

'My name is Hatteras,' I said. 'Mr Richard Hatteras.'

'Indeed, sir,' answered the lad. 'If you'll wait, Mr Dawson will see you in a minute, I'm sure.'

On hearing my name the other clerks began whispering together, at the same time throwing furtive glances in my direction.

In less than two minutes the clerk returned, and begged me to follow him, which I did. At the end of a long passage we passed through a curtained doorway, and I stood in the presence of the chief partner, Mr Dawson. He was a short, podgy man, with white whiskers and a bald head, and painfully precise.

'I have great pleasure in making your acquaintance, Mr Hatteras,' he said, as I came to an anchor in a chair. 'You have noticed our advertisement, I presume?'

'I saw it this morning,' I answered. 'And it is on that account I am here.'

'One moment before we proceed any further. Forgive what I am about to say—but you will see yourself that it is a point I am compelled not to neglect. Can you convince me as to your identity?'

'Very easily,' I replied, diving my hand into my breast-pocket and taking out some papers. 'First and foremost, here is my bank-book. Here is my card-case. And here are two or three letters addressed to me by London and Sydney firms. The Hon. Sylvester Wetherell, Colonial Secretary, will be glad, I'm sure, to vouch for me. Is that sufficient to convince you?'

'More than sufficient,' he answered, smiling. 'Now let me tell you for what purpose we desired you to call upon us.' Here he opened a drawer and took out a letter.

'First and foremost, you must understand that we are the Sydney agents of Messrs Atwin, Dobbs & Forsyth, of Furnival's Inn, London. From them, by the last English mail, we received this letter. I gather that you are the son of James Dymoke Hatteras, who was drowned at sea in the year 1880—is that so?'

'I am.'

'Your father was the third son of Sir Edward Hatteras of Murdlestone, in the county of Hampshire?'

'He was.'

'And the brother of Sir William, who had one daughter, Gwendoline Mary?'

'That is so.'

'Well, Mr Hatteras, it is my sad duty to inform you that within a week of your departure from England your cousin, the young lady just referred to, was drowned by accident in a pond near her home, and that her father, who had been ailing for some few days, died of heart disease on hearing the sad tidings. In that case, so my correspondents inform me, there being no nearer issue, you succeed to the title and estates—which I also learn are of considerable value, including the house and park, ten farms, and a large amount of house property, a rent roll of fifteen thousand a year, and accumulated capital of nearly a hundred thousand pounds.'

'Good gracious! Is this really true?'

'Quite true. You can examine the letter for yourself.'

I took it up from the table and read it through, hardly able to believe my eyes.

'You are indeed a man to be envied, Mr Hatteras,' said the lawyer. 'The title is an old one, and I believe the property is considered one of the best in that part of England.'

'It is! But I can hardly believe that it is really mine.'

'There is no doubt about that, however. You are a baronet as certainly as I am a lawyer. I presume you would like us to take whatever action is necessary in the matter?'

'By all means. This afternoon I am leaving Sydney, for a week or two, for the Islands. I will sign any papers when I come back.'

'I will bear that in mind. And your address in Sydney is——'

'Care of the Honourable Sylvester Wetherell, Potts Point.'

'Thank you. And, by the way, my correspondents have desired me on their behalf to pay in to your account at the Oceania the sum of five thousand pounds. This I will do today.'

'I am obliged to you. Now I think I must be going. To tell the truth, I hardly know whether I am standing on my head or my heels.'

'Oh, you will soon get over that.'

'Good-morning.'

'Good-morning, Sir Richard.'

With that, I bade him farewell, and went out of the office, feeling quite dazed by my good fortune. I thought of the poor idiot whose end had been so tragic, and of the old man as I had last seen him, shaking his fist at me from the window of the house. And to think that that lovely home was mine, and that I was a baronet, the principal representative of a race as old as any in the countryside! It seemed too wonderful to be true!

Hearty were the congratulations showered upon me at Pott Point, you may be sure, when I told my tale, and my health was drunk at lunch with much goodwill. But our minds were too much taken up with the arrangements for our departure that afternoon to allow us to think very much of anything else. By two o'clock we were ready to leave the house, by half-past we were on board the yacht, at three fifteen the anchor was up, and a few moments later we were ploughing our way down the harbour.

Our search for Phyllis had reached another stage.

CHAPTER V

The Islands and what we found there

To those who have had no experience of the South Pacific the constantly recurring beauties of our voyage would have seemed like a foretaste of Heaven itself. From Sydney, until the Loyalty Group lay behind us, we had one long spell of exquisite weather. By night under the winking stars, and by day in the warm sunlight, our trim little craft ploughed her way across smooth seas, and our only occupation was to promenade or loaf about the decks and to speculate as to the result of the expedition upon which we had embarked.

Having sighted the Isle of Pines we turned our bows almost due north and headed for the New Hebrides. Every hour our impatience was growing greater. In less than two days, all being well, we should be at our destination, and twenty-four hours after that, if our fortune proved in the ascendant, we ought to be on our way back with Phyllis in our possession once more. And what this would mean to me I can only leave you to guess.

One morning, just as the faint outline of the coast of Aneityum was peering up over the horizon ahead, Wetherell and I chanced to be sitting in the bows. The sea was as smooth as glass, and the tinkling of the water round the little vessel's nose as she turned it off in snowy lines from either bow, was the only sound to be heard. As usual the conversation, after wandering into other topics, came back to the subject nearest our hearts. This led us to make a few remarks about Nikola and his character. There was one thing I had always noticed when the man came under discussion, and that was the dread Wetherell had of him. My curiosity had been long excited as to its meaning, and, having an opportunity now, I could not help asking him for an explanation.

'You want to know how it is that I am so frightened of Nikola?' he asked, knocking the ash off his cigar on the upturned fluke of the anchor alongside him. 'Well, to give you my reason will necessitate my telling you a story. I don't mind doing that at all, but what I am afraid of is that you may be inclined to doubt its probability. I must confess it *is* certainly more like the plot of a Wilkie Collins novel than a bit of sober reality. However, if you want to hear it you shall.'

'I should like to above all things,' I replied, making myself comfortable and taking another cigar from my pocket. 'I have been longing to ask you about it for some time past, but could not quite screw up my courage.'

'Well, in the first place,' Mr Wetherell said, 'you must understand that before I became a Minister of the Crown, or indeed a Member of Parliament at all, I was a barrister with a fairly remunerative practice. That was before my wife's death and when Phyllis was at school. Up to the time I am going to tell you about I had taken part in no very sensational case. But my opportunity for earning notoriety was, though I did not know it, near at hand. One day I was briefed to defend a man accused of the murder of a Chinaman aboard a Sydney vessel on a voyage from Shanghai. At first there seemed to be no doubt at all as to his guilt, but by a singular chance, with the details of which I will not bore you, I hit upon a scheme which got him off. I remember the man perfectly, and a queer fellow he was, half-witted, I thought, and at the time of the trial within an ace of dying of consumption. His gratitude was the more pathetic because he had not the wherewithal to pay me. However, he made it up to me in another way, and that's where my real story commences.

'One wet night, a couple of months or so after the trial, I was sitting in my drawing-room listening to my wife's music, when a servant entered to tell me that a woman wanted to see me. I went out into the passage to find waiting there a tall buxom lass of about five-and-twenty years of age. She was poorly dressed, but in a great state of excitement.

' "Are you Mr Wetherell?" she said; "the gentleman as defended China Pete in the trial the other day?"

' "I am," I answered. "What can I do for you? I hope China Pete is not in trouble again?"

' "He's in a worse trouble this time, sir," said the woman. "He's dyin', and he sent me to fetch you to 'im before he goes."

' "But what does he want me for?" I asked rather suspiciously.

' "I'm sure I dunno," was the girl's reply. "But he's been callin' for you all this blessed day: 'Send for Mr Wetherell! send for Mr Wetherell!' So off I came, when I got back from work, to fetch you. If you're comin', sir, you'd best be quick, for he won't last till mornin'."

' "Very well, I'll come with you at once," I said, taking a mackintosh down from a peg as I spoke. Then, having told my wife not to sit up for me, I followed my strange messenger out of the house and down into the city.

'For nearly an hour we walked on and on, plunging deeper into the lower quarter of the town. All through the march my guide maintained a rigid silence, walking a few paces ahead, and only recognizing the fact that I was following her by nodding in a certain direction whenever we arrived at cross thoroughfares or interlacing lanes.

'At last we arrived at the street she wanted. At the corner she came suddenly to a standstill, and putting her two first fingers into her mouth blew a shrill whistle, after the fashion of street boys. A moment later a shock-headed urchin about ten years old made his appearance from a dark alley and came towards us. The woman said something to him, which I did not catch, and then turning sharply to her left hand beckoned to me to follow her. This I did, but not without a feeling of wonderment as to what the upshot of it all would be.

'From the street itself we passed, by way of a villainous alley, into a large courtyard, where brooded a silence like that of death. Indeed, a more weird and desolate place I don't remember ever to have met with. Not a soul was to be seen, and though it was surrounded by houses, only two feeble lights showed themselves. Towards one of these my guide made her way, stopping on the threshold. Upon a panel she rapped with her fingers, and as she did so a window on the first floor opened, and the same boy we had met in the street looked out.

' "How many?" enquired the woman, who had brought me, in a loud whisper.

' "None now," replied the boy; "but there's been a power of Chinkies hereabouts all the evenin', an' 'arf an hour ago there was a gent in a cloak."

'Without waiting to hear any more the woman entered the house and I followed close on her heels. The adventure was clearly coming to a head now.

'When the door had been closed behind us the boy appeared at the top of a flight of stairs with a lighted candle. We accordingly ascended to him, and having done so made our way towards a door at the end of the abominably dirty landing. At intervals I could hear the sound of coughing coming from a room at the end. My companion, however, bade me stop, while she went herself into the room, shutting the door after her. I was left alone with the boy, who immediately took me under his protection, and for my undivided benefit performed a series of highly meritorious acrobatic performances upon the feeble banisters, to his own danger, but apparent satisfaction. Suddenly, just as he was about to commence what promised to be the most successful item in his repertoire, he paused, lay flat on his stomach upon the floor, and craned his head over the side, where once banisters had been, and gazed into the half dark well below. All was quiet as the grave. Then, without warning, an almond-eyed, pigtailed head appeared on the stairs and looked upwards. Before I could say anything to stop him, the youth had divested himself of his one slipper, taken it in his right hand, leaned over a bit further, and struck the ascending Celestial a severe blow on the mouth with the heel of it. There was the noise of a hasty descent and the banging of the street door a moment later, then all was still again, and the youngster turned to me.

'"That was Ah Chong," he said confidentially. "He's the sixth Chinkie I've landed that way since dark."

'This important piece of information he closed with a double-jointed oath of remarkable atrocity, and, having done so, would have recommenced the performance of acrobatic feats had I not stopped him by asking the reason of his action. He looked at me with a grin, and said,—

'"I dunno, but all I cares is that China Pete in there gives me a sprat (sixpence) for every Chinkie what I keeps out of the 'ouse. He's a rum one is China Pete; an' can't he cough—my word!"

'I was about to put another question when the door opened and the girl who had brought me to the house beckoned me into the room. I entered and she left me alone with the occupant.

'Of all the filthy places I have ever seen—and I have had the ill-luck to discover a good many in my time—that one eclipsed them all. The room was at most ten feet long by seven wide, had a window at the far end, and the door, through which I had entered, opposite it. The bed-place was stretched between the door and the window, and was a horrible exhibition. On it, propped up by pillows and evidently in the last stage of collapse, was the man called China Pete, whom I had last seen walking out of the dock at the Supreme Court a couple of months before. When we were alone together he pointed to a box near the bed and signified that I should seat myself. I did so, at the same time taking occasion to express my sorrow at finding him in this lamentable condition. He made no reply to my civilities, but after a little pause found strength enough to whisper, "See if there's anybody at the door." I went across, opened the door and looked into the passage, but save the boy, who was now sitting on the top step of the stairs at the other end, there was not a soul in sight. I told him this, and having again closed the door, sat down on the box and waited for him to speak.

'"You did me a good turn, Mr Wetherell, over that trial," the invalid said at last, "and I couldn't make it worth your while."

'"Oh, you mustn't let that worry you," I answered soothingly. "You would have paid me if you had been able."

'"Perhaps I should, perhaps I shouldn't, anyhow I didn't, and I want to make it up to you now. Feel under my pillow and bring out what you find there."

'I did as he directed me and brought to light a queer little wooden stick about three and a half inches long, made of some heavy timber and covered all over with Chinese inscriptions; at one end was a tiny bit of heavy gold cord much tarnished. I gave it to him and he looked at it fondly.

'"Do you know the value of this little stick?" he asked after a while.

'"I have no possible notion," I replied.

'"Make a guess," he said.

'To humour him I guessed five pounds. He laughed with scorn.

'"Five pounds! O ye gods! Why, as a bit of stick it's not worth five pence, but for what it really is there is not money enough in the world to purchase it. If I could get about again I would make myself the richest and most powerful man on earth with it. If you could only guess one particle of the dangers I've been through to get it you would die of astonishment. And the sarcasm of it all is that now I've got it I can't make use of it. On six different occasions the priests of the Llamaserai in Peking have tried to murder me to get hold of it. I brought it down from the centre of China disguised as a wandering beggar. That business connected with the murder of the Chinaman on board the ship, against which you defended me, was on account of it. And now I lie here dying like a dog, with the key to over ten millions in my hand. Nikola has tried for five years to obtain it, without success however. He little dreams I've got it after all. If he did I'd be a dead man by this time."

' "Who is this Nikola then?" I asked.

' "Dr Nikola? Well, he's Nikola, and that's all I can tell you. If you're a wise man you'll want to know no more. Ask the Chinese mothers nursing their almond-eyed spawn in Peking who he is; ask the Japanese, ask the Malays, the Hindoos, the Burmese, the coal porters in Port Said, the Buddhist priests of Ceylon; ask the King of Corea, the men up in Thibet, the Spanish priests in Manilla, or the Sultan of Borneo, the ministers of Siam, or the French in Saigon— they'll all know Dr Nikola and his cat, and, take my word for it, they fear him."

'I looked at the little stick in my hand and wondered if the man had gone mad.

' "What do you wish me to do with this?" I asked.

' "Take it away with you," he answered, "and guard it like your life, and when you have occasion, use it. Remember you have in your hand what will raise a million men and the equivalent of over ten mil——"

'At this point a violent fit of coughing seized him and nearly tore him to pieces. I lifted him up a little in the bed, but before I could take my hands away a stream of blood had gushed from his lips. Like a flash of thought I ran to the door to call the girl, the boy on the stairs re-echoed my shout, and in less time than it takes to tell the woman was in the room. But we were too late—*China Pete was dead*.

'After giving her all the money I had about me to pay for the funeral, I bade her goodbye, and with the little stick in my pocket returned to my home. Once there I sat myself down in my study, took my legacy out of my pocket and carefully examined it. As to its peculiar power and value, as described to me by the dead man, I hardly knew what to think. My own private opinion was that China Pete was not sane at the time he told me. And yet, how was I to account for the affray with the Chinaman on the boat, and the evident desire the Celestials in Sydney had to obtain information concerning it? After half an hour's consideration of it I locked it up in a drawer of my safe and went upstairs to bed.

'Next day China Pete was buried, and by the end of the month I had well nigh forgotten that he had ever existed, and had hardly thought of his queer little gift, which still reposed in the upper drawer of my safe. But I was to hear more of it later on.

'One night, about a month after my coming into possession of the stick, my wife and I were entertaining a few friends at dinner. The ladies had retired to the drawing-room and I was sitting with the gentlemen at the table over our wine. Curiously enough we had just been discussing the main aspects of the politics of the East when a maid-servant entered to say that a gentleman had called, and would be glad to know if he might have an interview with me on important business. I replied to the effect that I was engaged, and told her to ask him if he would call again in the morning. The servant left the room only to return with the information that the man would be leaving Sydney shortly after daylight, but that if I would see him later on in the evening he would endeavour to return. I therefore told the girl to say I would see him about eleven o'clock, and then dismissed the matter from my mind.

'As the clock struck eleven I said goodnight to the last of my guests upon the doorstep. The carriage had not gone fifty yards down the street before a hansom drew up before my door and a man dressed in a heavy cloak jumped out. Bidding the driver wait for him he ran up my steps.

'"Mr Wetherell, I believe?" he said. I nodded and wished him "good-evening", at the same time asking his business.

'"I will tell you with pleasure," he answered, "if you will permit me five minutes alone with you. It is most important, and as I leave Sydney early tomorrow morning you will see that there is not much time to spare."

'I led the way into the house and to my study, which was in the rear, overlooking the garden. Once there I bade him be seated, taking up my position at my desk.

'Then, in the light of the lamp, I became aware of the extraordinary personality of my visitor. He was of middle height, but beautifully made. His face was oval in shape, with a deadly white complexion. In contrast to this, however, his eyes and hair were dark as night. He looked at me very searchingly for a moment and then said: "My business will surprise you a little I expect, Mr Wetherell. First, if you will allow me I will tell you something about myself and then ask you a question. You must understand that I am pretty well known as an Eastern traveller; from Port Said to the Kuriles there is hardly a place with which I am not acquainted. I have a hobby. I am a collector of Eastern curios, but there is one thing I have never been able to obtain."

'"And that is?"

'"A Chinese executioner's symbol of office."

'"But how can I help you in that direction?" I asked, completely mystified.

'"By selling me one that has lately come into your possession," he said. "It is a little black stick, about three inches long and covered with Chinese characters. I happened to hear, quite by chance, that you had one in your possession, and I have taken a journey of some thousands of miles to endeavour to purchase it from you."

'I went across to the safe, unlocked it, and took out the little stick China Pete had given me. When I turned round I almost dropped it with surprise as I saw the look of eagerness that rose in my visitor's face. But he pulled himself together and said, as calmly as he had yet addressed me:

'"That is the very thing. If you will allow me to purchase it, it will complete my collection. What value do you place upon it?"

'"I have no sort of notion of its worth," I answered, putting it down on the table and looking at it. Then in a flash a thought came into my brain, and I was about to speak when he addressed me again.

'"Of course my reason for wishing to buy it is rather a hare-brained one, but if you care to let me have it I will give you fifty pounds for it with pleasure."

'"Not enough, Dr Nikola," I said with a smile.

'He jumped as if he had been shot, and then clasped his hands tight on the arm of his chair. My random bolt had gone straight to the heart of the bullseye. This man then *was* Dr Nikola, the extraordinary individual against whom China Pete had warned me. I was determined now that, come what might, he should not have the stick.

'"Do you not consider the offer I make you a good one then, Mr Wetherell?" he asked.

'"I'm sorry to say I don't think the stick is for sale," I answered. "It was left to me by a man in return for a queer sort of service I rendered him, and I think I should like to keep it as a souvenir."

'"I will raise my offer to a hundred pounds in that case," said Nikola.

'"I would rather not part with it," I said, and as I spoke, as if to clinch the matter, I took it up and returned it to the safe, taking care to lock the door upon it.

'"I will give you five hundred pounds for it," cried Nikola, now thoroughly excited. "Surely that will tempt you?"

'"I'm afraid an offer of ten times that amount would make no difference," I replied, feeling more convinced than ever that I would not part with it.

'He laid himself back in his chair, and for nearly a minute and a half stared me full in the face. You have seen Nikola's eyes, so I needn't tell you what a queer effect they are able to produce. I could not withdraw mine from them, and I felt that if I did not make an effort I should soon be mesmerized. So, pulling myself together, I sprang from my chair, and, by doing so, let him see that our interview was at an end. However, he was not going without a last attempt to drive a bargain. When he saw that I was not to be moved his temper gave way, and he bluntly told me that I would *have* to sell it to him.

'"There is no compulsion in the matter," I said warmly. "The curio is my own property, and I will do just as I please with it."

'He thereupon begged my pardon, asked me to attribute his impatience to the collector's eagerness, and after a few last words bade me "goodnight", and left the house.

'When his cab had rolled away I went back to my study and sat thinking for awhile. Then something prompted me to take the stick out from the safe. I did so, and sat at my table gazing at it, wondering what the mystery might be to which it was the key. That it was not what Dr Nikola had described it I felt certain.

'At the end of half an hour I put it in my pocket, intending to take it upstairs to show my wife, locked the safe again and went off to my dressing-room. When I had described the interview and shown the stick to my wife I placed it in the drawer of the looking-glass and went to bed.

'Next morning, about three o'clock, I was awakened by the sound of some one knocking violently at my door. I jumped out of bed and enquired who it might be. To my intense surprise the answer was "Police!" I therefore donned my dressing-gown, and went out to find a sergeant of police on the landing waiting for me.

'"What is the matter?" I cried.

'"A burglar!" was his answer. "We've got him downstairs; caught him in the act."

'I followed the officer down to the study. What a scene was there! The safe had been forced, and its contents lay scattered in every direction. One drawer of my writing-table was wide open, and in a corner, handcuffed, and guarded by a stalwart constable, stood a Chinaman.

'Well, to make a long story short, the man was tried, and after denying all knowledge of Nikola—who, by the way, could not be found—was convicted, and sentenced to five years' hard labour. For a month I heard no more about the curio. Then a letter arrived from an English solicitor in Shanghai demanding from me, on behalf of a Chinaman residing in that place, a little wooden stick covered with Chinese characters, which was said to have been stolen by an Englishman, known in Shanghai as China Pete. This was very clearly another attempt on Nikola's part to obtain possession of it, so I replied to the effect that I could not entertain the request.

'A month or so later—I cannot, however, be particular as to the exact date—I found myself again in communication with Nikola, this time from South America. But there was this difference this time: he used undisguised threats, not only against myself, in the event of my still refusing to give him what he wanted, but also against my wife and daughter. I took no notice, with the result that my residence was again broken into, but still without success. Now I no longer locked the talisman up in the safe, but hid it in a place where I knew no one could possibly find it. My mind, you will see, was perfectly made up; I was not going to be driven into surrendering it.

'One night, a month after my wife's death, returning to my house I was garrotted and searched within a hundred yards of my own front door, but my assailants could not find it on me. Then peculiar pressure from other quarters was brought to bear; my servants were bribed, and my life became almost a burden to me. What was more, I began to develop that extraordinary fear of Nikola which seems to seize upon everyone who has any dealings with him. When I went home to England some months back, I did it because my spirits had got into such a depressed state that I could not remain in Australia. But I took care to deposit the stick with my plate in the bank before I left. There it remained till I returned, when I put it back in its old hiding-place again.

'The day after I reached London I happened to be crossing Trafalgar Square. Believing that I had left him at least ten thousand miles away, you may imagine my horror when I saw Dr Nikola watching me from the other side of the road. Then and there I returned to my hotel, bade Phyllis pack with all possible dispatch, and that same afternoon we started to return to Australia. The rest you know. Now what do you think of it all?'

'It's an extraordinary story. Where is the stick at the present moment?'

'In my pocket. Would you like to see it?'

'Very much, if you would permit me to do so.'

He unbuttoned his coat, and from a carefully contrived pocket under the arm drew out a little piece of wood of exactly the length and shape he had described. I took it from him and gazed at it carefully. It was covered all over with Chinese writing, and had a piece of gold silk attached to the handle. There was nothing very remarkable about it; but I must own I was strangely fascinated by it when I remembered the misery it had caused, the changes and chances it had brought about, the weird story told by China Pete, and the efforts that had been made by Nikola to obtain possession of it. I gave it back to its owner, and then stood looking out over the smooth sea, wondering where Phyllis was and what she was doing. Nikola, when I met him, would have a heavy account to settle with me, and if my darling reported any further cruelty on his part I would show no mercy. But why had Mr Wetherell brought the curio with him now? I put the question to him.

'For one very good reason,' he answered. 'If it is the stick Nikola is after, as I have every right to suppose, he may demand it as a ransom for my girl, and I am quite willing to let him have it. The wretched thing has caused sufficient misery to make me only too glad to be rid of it.'

'I hope, however, we shall be able to get her without giving it up,' I said. 'Now let us go aft to lunch.'

The day following we were within a hundred miles of our destination, and by midday of the day following that again were near enough to render it advisable to hold a council over our intended movements. Accordingly, a little before lunch-time the Marquis, Wetherell, the skipper and myself, met under the after awning to consider our plan of war. The vessel herself was hove to, for we had no desire to put in an appearance at the island during daylight.

'The first matter to be taken into consideration, I think, Mr Wetherell,' said the skipper, 'is the point as to which side of the island we shall bring up on.'

'You will be able to settle that,' answered Wetherell, looking at me. 'You are acquainted with the place, and can best advise us.'

'I will do so to the best of my ability,' I said, sitting down on the deck and drawing an outline with a piece of chalk. 'The island is shaped like this. There is no reef. Here is the best anchorage, without doubt, but here is the point where we shall be most likely to approach without being observed. The trend of the land is all upward from the shore, and, as far as I remember, the most likely spot for a hut, if they are detaining Miss Wetherell there, as we suppose, will be on a little plateau looking south, and hard by the only fresh water on the island.'

'And what sort of anchorage shall we get there, do you think?' asked the skipper, who very properly wished to run no risk with his owner's boat.

'Mostly coral. None too good, perhaps, but as we shall have steam up, quite safe enough.'

'And how do you propose that we shall reach the hut when we land? Is there any undergrowth, or must we climb the hill under the enemy's fire?'

'I have been thinking that out,' I said, 'and I have come to the conclusion that the best plan would be for us to approach the island after dark, to heave to about three miles out and pull ashore in the boat. We will then ascend the hill by the eastern slope and descend upon them. They will probably not expect us from that quarter, and it will at least be easier than climbing the hill in the face of a heavy fire. What do you say?'

They all agreed that it seemed practicable.

'Very good then,' said the skipper, 'we'll have lunch and afterwards begin our preparations.' Then turning to me, 'I'll get you to come into my cabin, Mr Hatteras, by-and-by and take a look at the Admiralty chart, if you will. You will be able probably to tell me if you think it can be relied on.'

'I'll do so with pleasure,' I answered, and then we went below.

Directly our meal was over I accompanied the skipper to look at the chart, and upon it we marked our anchorage. Then an adjournment was made aft, and our equipment of rifles and revolvers thoroughly overhauled. We had decided earlier that our landing party should consist of eight men—Wetherell, Beckenham, the mate of the yacht, myself, and four of the crew, each of whom would be supplied with a Winchester repeating rifle, a revolver, and a dozen cartridges. Not a shot was to be fired, however, unless absolutely necessary, and the greatest care was to be taken in order to approach the hut, if possible, without disturbing its inmates.

When the arms had been distributed and carefully examined, the sixteen foot surf-boat was uncovered and preparations made for hoisting her overboard. By the time this was done it was late in the afternoon, and almost soon enough for us to be thinking about overcoming the distance which separated us from our destination. Exactly at four o'clock the telegraph on the bridge signalled 'go ahead', and we were on our way once more. To tell the truth, I think we were all so nervous that we were only too thankful to be moving again.

About dusk I was standing aft, leaning against the taffrail, when Beckenham came up and stood beside me. It was wonderful what a difference these few months had made in him; he was now as brown as a berry, and as fine-looking a young fellow as any man could wish to see.

'We shall be picking up the island directly,' I said as he came to an anchor alongside me. 'Do you think you ought to go tonight? Remember you will run the risk of being shot!'

'I have thought of that,' he said. 'I believe it's my duty to do my best to help you and Mr Wetherell.'

'But what would your father say if he knew?'

'He would say that I only did what was right. I have just been writing to him, telling him everything. If anything *should* happen to me you will find the letter on the chest of drawers in your cabin. I know you will send it on to him. But if we both come out of it safely and rescue Miss Wetherell I'm going to ask a favour of you?'

'Granted before I know what it is!'

'It isn't a very big one. I want you to let me be your best man at your wedding?'

'So you shall. And a better I could not possibly desire.'

'I like to hear you say that. We've been through a good deal together since we left Europe, haven't we?'

'We have, and tonight will bring it to a climax, or I'm much mistaken.'

'Do you think Nikola will show fight?'

'Not a doubt about it I should think. If he finds himself cornered he'll probably fight like a demon.'

'It's Baxter I want to meet.'

'Nikola is my man. I've a big grudge against him, and I want to pay it.'

'How little we thought when we were cruising about Bournemouth Bay together that within such a short space of time we should be sailing the South Pacific on such an errand! It seems almost too strange to be possible.'

'So it does! All's well that ends well, however. Let's hope we're going to be successful tonight. Now I'm going on the bridge to see if I can pick the land up ahead.'

I left him and went forward to the captain's side. Dusk had quite fallen by this time, rendering it impossible to see very far ahead. A hand had been posted in the fore-rigging as a look-out, and every moment we expected to hear his warning cry; but nearly an hour passed, and still it did not come.

Then suddenly the shout rang out, 'Land ahead!' and we knew that our destination was in sight. Long before this all our lights had been obscured, and so, in the darkness—for a thick pall of cloud covered the sky—we crept up towards the coast. Within a couple of minutes of hearing the hail every man on board was on deck gazing ahead in the direction in which we were proceeding.

By tea-time we had brought the land considerably nearer, and by eight o'clock were within three miles of it. Not a sign, however, of any craft could we discover, and the greatest vigilance had to be exercised on our part to allow no sign to escape us to show our whereabouts to those ashore. Exactly at nine o'clock the shore party, fully armed, assembled on deck, and the surf-boat was swung overboard. Then in the darkness we crept down the gangway and took our places. The mate was in possession of the tiller, and when all was ready we set off for the shore.

CHAPTER VI
Conclusion

Once we had left her side and turned our boat's nose towards the land, the yacht lay behind us, a black mass nearly absorbed in the general shadow. Not a light showed itself, and everything was as still as the grave; the only noise to be heard was the steady dip, dip of the oars in the smooth water and now and then the chirp of the rowlocks. For nearly half an hour we pulled on, pausing at intervals to listen, but nothing of an alarming nature met our ears. The island was every moment growing larger, the beach more plain to the eye, and the hill more clearly defined.

As soon as the boat grounded we sprang out and, leaving one hand to look after her, made our way ashore. It was a strange experience that landing on a strange beach on such an errand and at such an hour, but we were all too much taken up with the work which lay before us to think of that. Having left the water's edge we came to a standstill beneath a group of palms and discussed the situation. As the command of the expedition had fallen upon me I decided upon the following course of action: To begin with, I would leave the party behind me and set out by myself to ascertain the whereabouts of the hut. Having discovered this I would return, and we would thereupon make our way inland and endeavour to capture it. I explained the idea in as few words as possible to my followers, and then, bidding them wait for me where they were, at the same time warning them against letting their presence be discovered, I set off up the hill in the direction I knew the plateau to lie. The undergrowth was very thick and the ground rocky; for this reason it was nearly twenty minutes before I reached the top of the hill. Then down the other side I crept, picking my way carefully, and taking infinite precautions that no noise should serve to warn our foes of my coming.

At last I reached the plateau and looked about me. A small perpendicular cliff, some sixty feet in height, was before me, so throwing myself down upon my stomach, I wriggled my way to its edge. When I got there I looked over and discovered three well-built huts on a little plateau at the cliff's base. At the same moment a roar of laughter greeted my ears from the building on the left. It was followed by the voice of a man singing to the accompaniment of a banjo. Under cover of his music I rose to my feet and crept back through the bushes, by the track along which I had come. I knew enough to distribute my forces now.

Having reached my friends again I informed them of what I had seen, and we then arranged the mode of attack as follows: The mate of the yacht, with two of the hands, would pass round the hill to the left of the plateau, Wetherell and another couple of men would take the right side, while Beckenham and myself crept down from the back. Not a sound was to be made or a shot fired until I blew my whistle. Then, with one last word of caution, we started on our climb.

By this time the clouds had cleared off the sky and the stars shone brightly. Now and again a bird would give a drowsy 'caw' as we disturbed him, or a wild pig would jump up with a grunt and go trotting off into the undergrowth, but beyond these things all was very still. Once more I arrived at the small precipice behind the huts, and, having done so, sat down for a few moments to give the other parties time to take up their positions. Then, signing to Beckenham to accompany me, I followed the trend of the precipice along till I discovered a place where we might descend in safety. In less than a minute we were on the plateau below, creeping towards the centre hut. Still our approach was undetected. Bidding Beckenham in a whisper wait for me, I crept cautiously round to the front, keeping as much as possible in the shadow. As soon as I had found the door, I tiptoed towards it and prepared to force my way inside, but I had an adventure in store for me which I had not anticipated.

Seated in the doorway, almost hidden in the shadow, was the figure of a man. He must have been asleep, for he did not become aware of my presence until I was within a foot of him. Then he sprang to his feet and was about to give the alarm. Before he could do so, however,

I was upon him. A desperate hand-to-hand struggle followed, in which I fought solely for his throat. This once obtained I tightened my fingers upon it and squeezed until he fell back unconscious. It was like a horrible nightmare, that combat without noise in the dark entry of the hut, and I was more than thankful that it ended so satisfactorily for me. As soon as I had disentangled myself, I rose to my feet and proceeded across his body into the hut itself. A swing door led from the porch, and this I pushed open.

'Who is it, and what do you want?' said a voice which I should have recognized everywhere.

In answer I took Phyllis in my arms and, whispering my name, kissed her over and over again. She uttered a little cry of astonishment and delight. Then, bidding her step quietly, I passed out into the starlight, leading her after me. As we were about to make for the path by which I had descended, Beckenham stepped forward, and at the same instant the man with whom I had been wrestling came to his senses and gave a shout of alarm. In an instant there was a noise of scurrying feet and a great shouting of orders.

'Make for the boats!' I cried at the top of my voice, and, taking Phyllis by the hand, set off as quickly as I could go up the path, Beckenham assisting her on the other side.

If I live to be a hundred I shall never forget that rush up the hill. In and out of trees and bushes, scratching ourselves and tearing our clothes, we dashed; conscious only of the necessity for speed. Before we were halfway down the other side Phyllis's strength was quite exhausted, so I took her in my arms and carried her the remainder of the distance. At last we reached the boats and jumped on board. The rest of the party were already there, and the word being given we prepared to row out to the yacht. But before we could push off a painful surprise was in store for us. The Marquis, who had been counting the party, cried:

'*Where is Mr Wetherell?*'

We looked round upon each other, and surely enough the old gentleman was missing. Discovering this, Phyllis nearly gave way and implored us to go back at once to find him. But having rescued her with so much difficulty I did not wish to run any risk of letting her fall into her enemies' hands again; so selecting four volunteers from the party, I bade the rest pull the boat out to the yacht and give Miss Wetherell into the captain's charge, while the balance accompanied me ashore again in search of her father. Having done this the boat was to return and wait for us.

Quickly we splashed our way back to the beach, and then, plunging into the undergrowth, began our search for the missing man. As we did not know where to search, it was like looking for a needle in a bundle of hay, but presently one of the hands remembered having seen him descending the hill, so we devoted our attentions to that side. For nearly two hours we toiled up and down, but without success. Not a sign of the old gentleman was to be seen. Could he have mistaken his way and be even now searching for us on another beach? To make sure of this we set off and thoroughly searched the two bays in the direction he would most likely have taken. But still without success. Perhaps he had been captured and carried back to the huts? In that case we had better proceed thither and try to rescue him. This, however, was a much more serious undertaking, and you may imagine it was with considerable care that we approached the plateau again.

When we reached it the huts were as quiet as when I had first made their acquaintance. Not a sound came up to the top of the little precipice save the rustling of the wind in the palms at its foot. It seemed difficult to believe that there had been such a tumult on the spot so short a time before.

Again with infinite care we crept down to the buildings, this time, however, without encountering a soul. The first was empty, so was the second, and so was the third. This result was quite unexpected, and rendered the situation even more mysterious than before.

By the time we had thoroughly explored the plateau and its surroundings it was nearly daylight, and still we had discovered no trace of the missing man. Just as the sun rose above the sea line we descended the hill again and commenced a second search along the beach, with no better luck, however, than on the previous occasion. Wetherell and our assailants seemed to have completely disappeared from the island.

About six o'clock, thoroughly worn out, we returned to the spot where the boat was waiting for us. What was to be done? We could not for obvious reasons leave the island and abandon the old gentleman to his fate, and yet it seemed useless to remain there looking for him, when he might have been spirited away elsewhere.

Conclusion

Suddenly one of the crew, who had been loitering behind, came into view waving something in his hand. As he approached we could see that it was a sheet of paper, and when he gave it into my hands I read as follows:

> If you cross the island to the north beach you will find a small cliff in which is a large cave, a little above high-water mark. There you will discover the man for whom you are searching.

There was no signature to this epistle, and the writing was quite unfamiliar to me, but I had no reason to doubt its authenticity.

'Where did you discover this?' I enquired of the man who had brought it.

'Fastened to one of them prickly bushes up on the beach there, sir,' he answered.

'Well, the only thing for us to do now is to set of to the north shore and hunt for the cave. Two of you had better take the boat back to the yacht and ask the captain to follow us round.'

As soon as the boat was under weigh we picked up our rifles and set off for the north beach. It was swelteringly hot by this time, and, as may be imagined, we were all dead tired after our long night's work. However, the men knew they would be amply rewarded if we could effect the rescue of the man for whom we had been searching, so they pushed on.

At last we turned the cape and entered the bay which constituted the north end of the island. It was not a large beach on this side, but it had, at its western end, a curious line of small cliffs, in the centre of which a small black spot could be discerned looking remarkably like the entrance to a cave. Towards this we pressed, forgetting our weariness in the excitement of the search.

It *was* a cave, and large one. So far the letter was correct.

Preparing ourselves, in case of surprise, we approached the entrance, calling Mr Wetherell's name. As our shouts died away a voice came out in answer, and thereupon we rushed in.

A remarkable sight met our eyes. In the centre of the cave was a stout upright post, some six or eight feet in height, and securely tied to this was the Colonial Secretary of New South Wales.

In less time almost than it takes to tell, we had cast loose the ropes which bound him, and led him, for he was too weak to stand alone, out into the open air. While he was resting he enquired after his daughter, and having learned that she was safe, gave us the following explanation. Addressing himself to me he said:

'When you cried "Make for the boats", I ran up the hill with the others as fast as I could go; but I'm an old man and could not get along as quickly as I wanted to, and for this reason was soon left far behind. I must have been half-way down the hill when a tall man, dressed in white, stepped out from behind a bush, and raising a rifle bade me come to a standstill. Having no time to lift my own weapon I was obliged to do as he ordered me, and he thereupon told me to lay down my weapon and right-about face. In this fashion I was marched back to the huts we had just left, and then, another man having joined my captor, was conducted across the island to this beach, where a boat was in waiting. In it I was pulled out to a small schooner lying at anchor in the bay and ordered to board her; five minutes later I was conducted to the saloon, where two or three persons were collected.

'"Good-evening, Mr Wetherell. This is indeed a pleasure," said a man sitting at the further end of the table. He was playing with a big black cat, and directly I heard his voice I knew that I was in the presence of Dr Nikola.

'"And how do you think I am going to punish you, my friend, for giving me all this trouble?" he said when I made no reply to his first remark.

'"You dare not do anything to me," I answered. "I demand that you let me go this instant. I have a big score to settle with you."

'"If you will be warned by me you will cease to demand," he answered, his eyes the while burning like coals. "You are an obstinate man, but though you have put me to so much trouble and expense I will forgive you and come to terms with you. Now listen to me. If you will give me——"

'At that moment the little vessel gave a heavy roll, and in trying to keep my footing on the sloping deck I fell over upon the table. As I did so the little Chinese stick slipped out of my pocket and went rolling along directly into Nikola's hands. He sprang forward and seized it, and you may imagine his delight. With a cry of triumph that made the cat leap from his shoulder, he turned to a tall man by his side and said:

'"I've got it at last! Now let a boat's crew take this man ashore and tie him to the stake in the cave. Then devise some means of acquaint-

ing his friends of his whereabouts. Be quick, for we sail in an hour."
Having given these orders he turned to me again and said:

' "Mr Wetherell, this is the last transaction we shall probably ever have together. All things considered, you are lucky in escaping so easily. It would have saved you a good deal if you had complied with my request at first. However, all's well that ends well, and I congratulate you upon your charming daughter. Now, goodbye; in an hour I am off to effect a *coup* with this stick, the magnitude of which you would never dream. One last word of advice: pause a second time, I entreat, before you think of baulking Dr Nikola."

'I was going to reply, when I was twisted round and led up on deck, where that scoundrel Baxter had the impudence to make me a low bow. In less than a quarter of an hour I was fastened to the post in that cave. The rest you know. Now let us get on board; I see the boat is approaching.'

As soon as the surf-boat had drawn up on the beach we embarked and were pulled out to the yacht. In a few moments we were on deck, and Phyllis was in her father's arms again. Over that meeting, with its rapturous embraces and general congratulations, I must draw a curtain. Suffice it that by midday the island had disappeared under the sea line, and by nightfall we were well on our way back to Sydney.

That evening, after dinner, Phyllis and I patrolled the deck together, and finally came to a standstill aft. It was as beautiful an evening as any man or woman could desire. All round us was the glassy sea, rising and falling as if asleep, while overhead the tropic stars shone down with their wonderful brilliance.

'Phyllis,' I said, taking my darling's hand in mine and looking into her face, 'what a series of adventures we have both passed through since that afternoon I first saw you in the Domain! Do you know that your father has at last consented to our marriage?'

'I do. And as it is to you, Dick, I owe my rescue,' she said, coming a little closer to me, 'he could do nothing else; you have a perfect right to me.'

'I have, and I mean to assert it!' I answered. 'If I had not found you, I should never have been happy again.'

'But, Dick, there is one thing I don't at all understand. At dinner this evening the captain addressed you as Sir Richard. What does that mean?'

'Why, of course you have not heard!' I cried. 'Well, I think it means that though I cannot make you a marchioness, I can make you a baronet's wife. It remains with you to say whether you will be Lady Hatteras or not.'

'But are you a baronet, Dick? How did that come about?'

'It's a long story, but do you remember my describing to you the strange call I paid, when in England, on my only two relatives in the world?'

'The old man and his daughter in the New Forest? Yes, I remember.'

'Well, they are both dead, and, as the next-of-kin, I have inherited the title and estates. What do you think of that?'

Her only reply was to kiss me softly on the cheek.

She had scarcely done so before her father and Beckenham came along the deck.

'Now, Phyllis,' said the former, leading her to a seat, 'supposing you give us the history of your adventures. Remember we have heard nothing yet.'

'Very well. Where shall I begin? At the moment I left the house for the ball? Very good. Well, you must know that when I arrived at Government House I met Mrs Mayford—the lady who had promised to chaperone me—in the cloakroom, and we passed into the ballroom together. I danced the first dance with Captain Hackworth, one of the aides, and engaged myself for the fourth to the Marquis of Beckenham.'

'The sham Marquis, unfortunately,' put in the real one.

'It proved to be unfortunate for me also,' continued Phyllis. 'As it was a square we sat it out in the ante-room leading off the drawing-room, and while we were there the young gentleman did me the honour of proposing to me. It was terribly embarrassing for me, but I allowed him to see, as unmistakably as possible, that I could give him no encouragement, and, as the introduction to the next waltz started, we parted the best of friends. About half an hour later, just as I was going to dance the lancers, Mrs Mayford came towards me and drew me into the drawing-room. Mr Baxter, his lordship's tutor, was with her, and I noticed that they both looked supernaturally grave.

' "What is the matter?" I asked, becoming alarmed by her face.

' "My dear," said she, "you must be brave. I have come to tell you that your father has been taken ill, and has sent for you."

' "Papa ill!" I cried. "Oh, I must go home to him at once."

' "I have taken the liberty of facilitating that," said Mr Baxter, "by ordering the servants to call up your carriage, which is now waiting for you at the door. If you will allow me, I will conduct you to it?"

'I apologized to my partner for being compelled to leave him, and then went to the cloakroom. As soon as I was ready I accompanied Mr Baxter to the door, where the brougham was waiting. Without looking at the coachman I got in, at the same time thanking my escort for his kindness. He shut the door and cried "Home" to the coachman. Next moment we were spinning down the drive.

'As I was far too much occupied thinking of you, papa, I did not notice the direction we were taking, and it was not until the carriage stopped before a house in a back street that I realized that something was wrong. Then the door was opened and a gentleman in evening dress begged me to alight. I did so, almost without thinking what I was doing.

' "I am sorry to say your father is not at all well, Miss Wetherell," said the person who helped me out. 'If you will be good enough to step into my house I will let the nurse take you to him."

'Like a person in a dream I followed him into the dwelling, and, as soon as I was inside, the door was shut upon me.

'Where is my father? and how is it that he is here?' I cried, beginning to get frightened.

' "You will know all when you see him," said my companion, throwing open the door of a bedroom. I went in, and that door was also shut upon me. Then I turned and faced the man.'

'What was he like?' cried Wetherell.

'He was the man you were telling us about at dinner—Dr Nikola.'

'Ah! And then?'

'He politely but firmly informed me that I was his prisoner, and that until you gave up something he had for years been trying to obtain he would be compelled to detain me. I threatened, entreated, and finally wept, but he was not to be moved. He promised that no effort should be spared to make me comfortable, but he could not let me go until you had complied with his request. So I was kept there until late one night, when I was informed that I must be ready to leave the house. A brougham was at the door, and in this, securely guarded, I was conducted to the harbour, where a boat was in waiting. In this we were rowed out to a schooner, and I was placed on board her. A comfortably furnished cabin was allotted to me, and everything I could possibly want was given me. But though the greatest consideration in all other matters was shown me I could gather nothing of where we were going or what my fate was to be, nor could I discover any means of communicating with the shore. About midnight we got under weigh and commenced our voyage. Our destination was the island where you found me.'

'And how did Nikola treat you during the voyage and your stay on Pipa Lannu?' I asked.

'With invariable courtesy,' she replied. 'A more admirable host no one could desire. I had but to express a wish and it was instantly gratified. When we were clear of the land I was allowed on deck; my meals were served to me in a cabin adjoining my own, and a stewardess had been specially engaged to wait upon me. As far as my own personal treatment went I have nothing to complain of. But oh, you can't tell how thankful I was to get away; I had begun to imagine all sorts of horrors.'

'Well, God be thanked, it's all done with now,' I said earnestly.

'And what is more,' said Wetherell, 'you have won one of the best husbands in the world. Mr Hatteras, your hand, sir; Phyllis, my darling, yours! God bless you both.'

Now what more is there to tell? A week later the eventful voyage was over and we were back in Sydney again.

Then came our marriage. But, with your kind permission, I will only give you a very bare description of that. It took place at the cathedral, the Primate officiating. The Marquis of Beckenham was kind enough to act as my best man, while the Colonial Secretary, of course, gave his daughter away.

But now I come to think of it, there is one point I *must* touch upon in connection with that happy occasion, and that was the arrival of an important present on the evening prior to the event.

We were sitting in the drawing-room when the butler brought in a square parcel on a salver and handed it to Phyllis.

'Another present, I expect,' she said, and began to untie the string that bound it.

When the first cover was removed a layer of tissue paper revealed itself, and after that a large Russia leather case came into view. On pressing the spring the cover lifted and revealed a superb collet—as

I believe it is called—of diamonds, and resting against the lid a small card bearing this inscription:

> With heartiest congratulations and best wishes to Lady Hatteras, in memory of an unfortunate detention and a voyage to the Southern Seas,
> From her sincere admirer,
> Dr Nikola.

What do you think of that?

Well, to bring my long story to a close, the Great Event passed off with much éclat. We spent our honeymoon in the Blue Mountains, and a fortnight later sailed once more for England in the *Orizaba*. Both Mr Wetherell—who has now resigned office—and the Marquis of Beckenham, who is as manly a fellow as you would meet anywhere in England, accompanied us home, and it was to the latter's seaside residence that we went immediately on our arrival in the mother country. My own New Forest residence is being thoroughly renovated, and will be ready for occupation in the spring.

And now as to the other persons who have figured most prominently in my narrative. Of Nikola, Baxter, Eastover, or Prendergast I have never heard since. What gigantic coup the first-named intends to accomplish with the little Chinese stick, the possession of which proved so fatal to Wetherell, is beyond my power to tell. I am only too thankful, however, that I am able to say that I am not in the least concerned in it. I am afraid of Nikola and I confess it. And with this honest expression of my feelings, and my thanks for your attention and forbearance, I will beg your permission to ring the curtain down upon the narrative of my BID FOR FORTUNE.

MORE OXFORD PAPERBACKS

This book is just one of nearly 1000 Oxford Paperbacks currently in print. If you would like details of other Oxford Paperbacks, including titles in the World's Classics, Oxford Reference, Oxford Books, OPUS, Past Masters, Oxford Authors, and Oxford Shakespeare series, please write to:

UK and Europe: Oxford Paperbacks Publicity Manager, Arts and Reference Publicity Department, Oxford University Press, Walton Street, Oxford OX2 6DP.

Customers in UK and Europe will find Oxford Paperbacks available in all good bookshops. But in case of difficulty please send orders to the Cash-with-Order Department, Oxford University Press Distribution Services, Saxon Way West, Corby, Northants NN18 9ES. Tel: 01536 741519; Fax: 01536 746337. Please send a cheque for the total cost of the books, plus £1.75 postage and packing for orders under £20; £2.75 for orders over £20. Customers outside the UK should add 10% of the cost of the books for postage and packing.

USA: Oxford Paperbacks Marketing Manager, Oxford University Press, Inc., 200 Madison Avenue, New York, N.Y. 10016.

Canada: Trade Department, Oxford University Press, 70 Wynford Drive, Don Mills, Ontario M3C 1J9.

Australia: Trade Marketing Manager, Oxford University Press, G.P.O. Box 2784Y, Melbourne 3001, Victoria.

South Africa: Oxford University Press, P.O. Box 1141, Cape Town 8000.

THE CONCISE OXFORD COMPANION TO ENGLISH LITERATURE

Edited by Margaret Drabble and Jenny Stringer

Derived from the acclaimed *Oxford Companion to English Literature*, the concise maintains the wide coverage of its parent volume. It is an indispensable, compact guide to all aspects of English literature. For this revised edition, existing entries have been fully updated and revised with 60 new entries added on contemporary writers.

* **Over 5,000 entries on the lives and works of authors, poets and playwrights**
* **The most comprehensive and authoritative paperback guide to English literature**
* **New entries include Peter Ackroyd, Martin Amis, Toni Morrison, and Jeanette Winterson**
* **New appendices list major literary prize-winners**

From the reviews of its parent volume:

'It earns its place at the head of the best sellers: every home should have one'
Sunday Times

ILLUSTRATED HISTORIES IN OXFORD PAPERBACKS

THE OXFORD ILLUSTRATED HISTORY OF ENGLISH LITERATURE

Edited by Pat Rogers

Britain possesses a literary heritage which is almost unrivalled in the Western world. In this volume, the richness, diversity, and continuity of that tradition are explored by a group of Britain's foremost literary scholars.

Chapter by chapter the authors trace the history of English literature, from its first stirrings in Anglo-Saxon poetry to the present day. At its heart towers the figure of Shakespeare, who is accorded a special chapter to himself. Other major figures such as Chaucer, Milton, Donne, Wordsworth, Dickens, Eliot, and Auden are treated in depth, and the story is brought up to date with discussion of living authors such as Seamus Heaney and Edward Bond.

'[a] lovely volume . . . put in your thumb and pull out plums' Michael Foot

'scholarly and enthusiastic people have written inspiring essays that induce an eagerness in their readers to return to the writers they admire' *Economist*

PAST MASTERS

PAST MASTERS

A wide range of unique, short, clear introductions to the lives and work of the world's most influential thinkers. Written by experts, they cover the history of ideas from Aristotle to Wittgenstein. Readers need no previous knowledge of the subject, so they are ideal for students and general readers alike.

Each book takes as its main focus the thought and work of its subject. There is a short section on the life and a final chapter on the legacy and influence of the thinker. A section of further reading helps in further research.

The series continues to grow, and future Past Masters will include **Owen Gingerich** on *Copernicus*, **R G Frey** on *Joseph Butler*, **Bhiku Parekh** on *Gandhi*, **Christopher Taylor** on *Socrates*, **Michael Inwood** on *Heidegger*, and **Peter Ghosh** on *Weber*.